OUR LIVING WORLD OF NATURE

# *The Life of the Jungle*

*Developed jointly with The World Book Encyclopedia*

*Produced with the cooperation of The United States Department of the Interior*

OUR LIVING WORLD OF NATURE

# The Life of the Jungle

PAUL W. RICHARDS



*Published in cooperation with*
*The World Book Encyclopedia*
*McGraw-Hill Book Company*
NEW YORK TORONTO LONDON

**PAUL W. RICHARDS** *is Professor of Botany at the University College of North Wales in Bangor, Wales, United Kingdom. He was educated at Cambridge University's Trinity College, where he was awarded the B.A., M.A., Ph.D., and Sc.D. degrees. After teaching botany for several years at Cambridge, he joined the faculty at the University College of North Wales and was for many years Head of the Department of Botany there. He has also lectured at universities in Nigeria, Finland, Belgium, and the United States. Dr. Richards has served as president of both the British Bryological Society and the British Ecological Society and has been editor of the* Journal of Ecology. *He recalls that his earliest interest in the tropics was aroused by reading the classic works of the great British naturalists, Charles Darwin, Alfred Russel Wallace, and Henry Walter Bates. He first visited a tropical rain forest as a member of the Oxford University Expedition to British Guiana in 1929, while still an undergraduate at Cambridge. He has since conducted research in many of the tropical rain forests of both the Old and the New Worlds, most recently in Brazil and Ghana. Dr. Richards is the author of* The Tropical Rain Forest, A Book of Mosses, *and numerous scientific papers.*



Library of Congress Catalog Card Number: 70–103911

*12345678910 NR RM 76543210*

46012

## OUR LIVING WORLD OF NATURE

*Science Editor*

| | |
|---|---|
| RICHARD B. FISCHER | *Cornell University* |

*Board of Consultants*

| | |
|---|---|
| ROLAND CLEMENT | *National Audubon Society* |
| C. GORDON FREDINE | *National Park Service, The United States Department of the Interior* |
| WILLIAM H. NAULT | *Field Enterprises Educational Corporation* |
| BENJAMIN NICHOLS | *Cornell University* |
| OLIN SEWALL PETTINGILL, JR. | *Cornell University* |
| DAVID PIMENTEL | *Cornell University* |
| EDGAR M. REILLY, JR. | *New York State Museum* |
| PAUL B. SEARS | *Yale University* |

*Readability Consultant*

| | |
|---|---|
| JOSEPHINE PIEKARZ IVES | *New York University* |

*Special Consultants for The Life of the Jungle*

| | |
|---|---|
| F. RAYMOND FOSBERG | *Smithsonian Institution* |
| A. J. CAIN | *University of Liverpool* |

# Contents

# *The Jungle World*

The river is your highway into the jungle, and with each stroke of the paddle you draw nearer to your destination. But as the canoe glides swiftly across the dark coffee-colored water, you probably wonder how you will make your way into the dim mysterious world of the tropical forest. The question is a good one, for on either bank the river is lined by a wall of vegetation, one hundred feet high perhaps, and so dense that you can see nothing beyond. All the stories you have heard about the impenetrable jungle seem to be true, and you imagine yourself slashing a path through hopeless snarls of brush and vines.

Suddenly the native guide turns the canoe toward shore. There is a small dark gap in the curtain of foliage, but it is not until the boat is almost ready to enter the opening that you realize it is the mouth of a side stream. Once the boat enters the side stream, it begins to progress more slowly, for the creek is narrow and winding. Fallen logs and overhanging branches frequently block the way. As the stream grows narrower, the trees begin to meet overhead so that you see less and less of the sky. On the river the boat seemed to be passing between the walls of a broad green canyon, but on the stream you feel as if you are in a tunnel.

Soon you come to the place where your guide will set up camp, a well-drained sandy slope beside the creek. Taking your gear ashore presents no great problem, for here where little sunlight reaches the forest floor the undergrowth is rather thin. Much of it consists of saplings with stems no thicker than your little finger. Now and then your path is blocked by a fallen tree or a tangle of vines, and occasionally you have to push aside a low-hanging branch. But you can walk long distances without using a machete to clear a path. It is so open, in fact, that you can easily see a companion sixty yards away.

As you look around, you realize that tales about the impenetrable jungle are mostly myths. It is true that where the original jungle has been cleared away, the second-growth forest that springs up in its place *is* nearly impenetrable for a time as a profusion of plants reaches for the sun. But most of the stories arose because early explorers usually stayed on the rivers, where travel was easiest. Since the rivers form openings where the sunlight can reach all the way to the

**Its bristly crest readily identifies a hoatzin peering curiously from a leafy perch beside a jungle stream. These strange, two-foot-long birds live in small flocks near swampy forests and jungle streams in South America.**

ground, they are always lined by curtains of greenery. The same is true of any clearing in the jungle. But a short distance from the edge of the clearing, the undergrowth thins out. It is simply too shady for many plants to grow at ground level beneath the tall jungle trees.

### *A jungle bird*

Your camp is in the heart of Guyana, on the northern coast of South America. It is a good place to begin your exploration of the jungle, or *tropical rain forest,* as botanists prefer to call it. Other great jungles straddle the equator in West Africa, southeast Asia, and a few other places, but the South American jungle is the largest in the world.

You are especially fortunate because your first glimpse of jungle life is one of the specialties of the South American rain forest. Your guide leads you to a swampy thicket near the stream and there lurking in the trees are several slender dark brown birds, each one about two feet long. Their most notable features are their long, bristly, reddish-brown crests,

**A pair of movable claws on each wing enables young hoatzins to clamber from branch to branch soon after hatching. The claws drop off within two or three weeks, but even older young like this one continue to use their wings as hooks to assist in climbing.**

which give the birds a rather comical look. Although they hiss loudly at your approach and flutter awkwardly to higher perches, they do not seem particularly alarmed. The birds are hoatzins, and they live only in the jungles and along settled river banks in northern South America.

The hoatzins' young are especially interesting. The birds usually build their flimsy nests on branches that overhang the water. In times of danger, the little birds plunge overboard without hesitation and swim to safety in the tangled vegetation along the shore. Even more astonishing is their ability to climb. Unlike most birds, they have two movable claws near the tip of each wing. They use the claws as hooks to haul themselves from twig to twig. Although they lose the claws as they mature, even adult hoatzins climb nearly as often as they fly, using their wings like hands as they clamber through the treetops.

## *A world of trees*

But aside from these birds, you notice few signs of animal life at first. Now and then a butterfly flutters past. It may be a *Heliconius*, with long, narrow, jet-black wings streaked and spotted with brilliant crimson and yellow. Or perhaps by the stream you see a blue morpho, with large incredibly blue wings that glitter like metal in the sunlight. Occasionally you hear the curious whistle of the greenheart bird, one of the many kinds of cotingas that live in the American tropics. The most extraordinary of the cotingas are the bellbirds, whose strange bell-like calls are so loud and clear that they can be heard more than a mile away.

But, except for the sounds of insects, the jungle is silent, with not even a breath of wind to rustle the leaves. And everywhere you look there are trees. Trees, and more trees—from little saplings of less than a man's height to giants 150 feet high or even taller. Although the trees are tall, it is surprising to notice how few really large trunks there are. None

**Silhouetted against the misty light in a jungle glade in northern South America, the straight slender trunks of giant trees rise 100 feet or more toward the sky. In some places, the tallest rain-forest trees may grow 150 or even 200 feet tall.**

Many of the larger jungle trees are rimmed by buttresses, thin triangular plates of very hard wood that radiate from the bases of their trunks. The buttresses act as braces that help support tall, shallow-rooted trees and prevent them from being toppled by strong winds.

of them even approaches the girth of California's famous sequoias. Here and there is a tree three feet or more in diameter, but the vast majority of the trunks are a foot or less across.

At first glance the trees all look alike. Most of them have smooth, rather light-colored bark, and dark green leathery leaves. There are differences, of course, but it takes a keen eye to notice them. A skilled tree finder can show you how to identify trees by clues such as subtle variations in the color and texture of the bark or even differences in the taste and smell of the wood.

Since their leaves and bark all tend to look alike, it is usually difficult to identify jungle trees. But slashing the trunks with a machete often provides useful clues. Thick, dull-reddish bark and drops of yellow latex oozing from the cut identify this tree as the African mammee apple. This useful timber tree, which grows to twelve feet in girth, is common in West African jungles.

As the guide shows you some of these differences, you begin to realize that literally hundreds of kinds of trees grow in the forest near your camp. In the jungle, it is not at all unusual to find over sixty different species a foot or more in diameter on a single acre of land. Compare this with the figures for familiar temperate forests, and you will begin to appreciate the profusion of jungle life. An acre of woodland in England is likely to include only three or four species of trees. In the Appalachian forests of the United States—probably the richest in tree species of any temperate-zone forests—you would find no more than about twenty-five species of trees on an acre of woodland.

Even more surprising are the botanical relationships of some of the jungle trees. In tropical forests many large trees belong to families that in temperate zones include mostly small soft-stemmed plants. In the jungle, many relatives of our familiar violets, for example, are trees that grow thirty feet tall or even higher. Similarly, most of the milkworts found in the United States are little herbaceous plants, but one of the milkworts of the Amazon jungle is a large tree with very hard wood.

Another striking feature of many of the trees is the *buttresses* that radiate from their bases. These are thin, triangular plates of very hard wood that form in the angles between the trunk and horizontal roots running near the surface of the ground. The buttresses—usually three or four to a tree but sometimes as many as ten—may extend six feet out from the base of the tree and six feet or more up the trunk. As a result, lumbermen often have to build platforms around large trees so that they can cut the trunks above the buttresses.

Although some species of trees always have buttresses, others never do. Somewhat similar are the *prop roots* found on many other kinds of trees. These roots, which curve out from the trunk a few feet above the ground, make the trees look as if they are standing on stilts. Prop roots and buttresses are a common sight in jungles throughout the world, but, curiously, they are seldom found on trees outside the tropics. Some scientists suggest that they are useful to shallow-rooted jungle trees as stays or braces that help the trees resist overthrow by the wind. But, if so, why have not shallow-rooted temperate-zone trees evolved similar supports?

## *Plants underfoot, plants overhead*

With so much foliage overhead, it is hardly surprising that the jungle floor is shady. Yet it is not the place of perpetual gloom that some writers picture. As long as the sun is high in the sky, shafts of light here and there penetrate gaps in the foliage and dapple the jungle floor with sunflecks. In the morning and evening when the sun is low, however, little light penetrates the treetops, so that the day in the forest is perhaps two hours shorter than in the open.

**Other than the seedlings and saplings of trees and vines, relatively few plants grow in the dim light beneath rain-forest trees. But where there is a gap in the canopy, bushes such as this *Cephaelis* of the Amazon jungle find enough light to grow.**

Despite this relative shortage of sunlight, quite a variety of plants are able to survive on the ground beneath jungle trees and perched on the trees themselves. It is not like a temperate-zone forest, of course. There are relatively few flowering plants such as the trilliums, mayflowers, and violets that carpet the ground in a New England woodland. But many kinds of plants, including seedlings of the trees themselves, are able to thrive on the meager ration of sunlight provided by the flecks of sun that pass across their leaves each day. Ferns, mosses, liverworts, and lichens grow not only on the ground and on rotting logs, but even on the trunks of living trees. There are also palms of various sizes —some with fiercely thorny stems—but they grow mainly in small openings and by streams.

All in all, however, the undergrowth is rather sparse except where gaps in the foliage let in more light. But overhead is a whole new world of plant life. In some places, practically every tree seems to provide support for one or more vines. Some are thin and, like English ivy, cling to the tree trunks by means of tiny, so-called *aerial roots* that grow along their stems.

But the most spectacular are the great woody vines called *lianas*, a characteristic feature of jungle vegetation everywhere. The larger ones are as thick as a man's arm or thigh; rarely they may be as much as two feet in diameter. The stems of some lianas are flattened like enormous ribbons; others are twisted like ropes or wire cables.

In some places, nearly every trunk seems to support one or more lianas, which disappear in the foliage overhead. In the treetops, the lianas form fantastic interlacing networks and loop like giant serpents from tree to tree. They form such tangled networks, in fact, that they will often hold a tree up even after its base has been cut.

Lianas have solved the problem of survival in the dimly lit jungle by escaping to the sunny world overhead. Their stems shoot up rapidly, using the tree trunks for support. When they finally reach the treetops, they are able to live in full sunlight.

Another type of plant also solves the problem of getting adequate sunlight by living in the treetops. These are the *epiphytes,* or air plants. Like lianas, they are characteristic plant types in the tropical rain forest. Their name derives from two Greek words: *epi,* meaning "upon," and *phyton,* meaning "plant." And they are indeed plants that live upon other plants, far away from the soil. Some of them grow from cracks and crevices in the bark on tree trunks, and even more are perched on the upper sides of branches.

Most of the epiphytes are too high and too well hidden by the foliage to be visible from below. The only way to study them from the ground may be to find a gap in the canopy of leaves and peer at the treetops through your binoculars.

But occasionally you have an opportunity to examine them at close hand. From time to time some of the plants slip from their attachment up above and continue to live for awhile in the much shadier conditions at ground level. Or an entire branch may snap off in a storm and come crashing down, bringing a whole cargo of epiphytes with it.

Many of the epiphytes are small ferns, but others are flowering plants of several kinds. Some are orchids, with blossoms of many shapes, sizes, and colors. Others, sometimes more showy than the orchids, are bromeliads, members of the pineapple family. As you look at them and marvel at the variety of their forms, you will realize that there is indeed

**Much more numerous than the herbaceous plants of the forest floor are the many kinds of epiphytes that perch on the twigs and branches of jungle trees. Epiphytes are not parasitic; they manufacture their own food and depend on the host trees only for support.**

Henri Julien Roussea

a hidden world in the treetops. Less than thirty or forty feet above ground level, flowers are few and far between, but the treetops, it seems, are full of color. As a matter of fact, in the tropical rain forest many more herbaceous plant species grow as epiphytes than on the ground. Later we will learn more about the ways in which they cope with the special problems of living up in the air on other plants.

### *Misleading myths*

From our first glimpse of the tropical rain forest, it is clear that many popular notions about the jungle have little to do with the facts. A mature rain forest that has never been felled or logged is not impenetrable by any means. The light may be dim, but when the sun is shining the forest certainly is not gloomy. And the jungle at ground level is not particularly colorful. The traveler's first impression, in fact, is one of muted browns and greens. There are some blossoms, but many of them are tiny green or white flowers that only a botanist would notice. And the more colorful epiphytes, as we have just seen, are hidden in the foliage overhead.

The same is true of the trees and vines. Many of them bear showy flowers, but most of the blossoms are so high up that you can glimpse them only where there is a break in the canopy. More often you can tell that a tree is flowering only by the scattering of fallen petals at its base or by clues such as the humming of bees or a wave of strong sweet scent. And frequently the masses of bright red, white, or purple that you glimpse overhead prove not to be flowers at all but branches bursting into leaf instead: it is common for the young leaves of jungle trees to be brightly colored.

Another common misconception about the jungle is that the ground is covered with an age-long accumulation of decaying vegetation. In fact, the leaf litter in many places is hardly enough to cover the bare soil. The only sizable accumulations are in hollows and between the buttresses at the bases of trees.

**When the French primitive painter Henri Rousseau painted his "Equatorial Jungle," he combined his own recollections of the Mexican jungle, where he had served as a soldier, with careful observation of tropical plants in the Paris botanical garden.**

There is good reason for this. Under the hot, moist conditions of the humid tropics, the organisms that destroy dead plant material work far more rapidly than in cooler climates. Every breath of wind brings dead leaves, twigs, and fruits to the ground. But ants, termites, mites, worms, fungi, bacteria, and a host of other organisms break them down so quickly that little dead material accumulates.

### *Where are the animals?*

But what about animals, you wonder? Certainly the stories you have heard about the profusion of animal life in the jungle must be true. Surely there are eyes peering from beneath every leaf and great beasts lurking behind every tree trunk.

In this case the tales you may have heard are not exactly false. But they are a little misleading. Although the jungle is alive with animals, many of the creatures are extremely difficult to see. Most of the larger animals are very shy and will hide at your approach. Others are active only at night. And many of them spend their entire lives high up in the treetops.

**Several species of curassows, all of them quite similar in appearance, are found in the South American rain forest. These large turkeylike birds feed in groups on the jungle floor, but they roost and nest in the lower branches of trees.**

The most conspicuous and by far the most numerous animals are insects. At Kartabo in Guyana, the famous naturalist William Beebe once tried to record in his notebook every living creature he saw in a single hour. From 7:30 to 8:15 in the morning he walked slowly through the jungle, then remained motionless for the rest of the hour. During this time he saw 636 individual animals. Twelve of them were mammals, 128 were birds, 13 were spiders, and 113 were animals of various other types. But the vast majority—370 of the 636, to be exact—were insects.

Insects, in fact, seem to be everywhere. Countless species of butterflies, flies, beetles, bugs, mantids, grasshoppers, bees, wasps, and every other type of insect live at all levels, from the ground to the treetops, and underground as well. Their subdued droning, humming, and screeching is so nearly continuous that before long you cease to notice the sound. You are likely to pay attention only when the music suddenly changes. Each time the sun comes from behind a cloud, for instance, the cicadas burst into a deafening chorus.

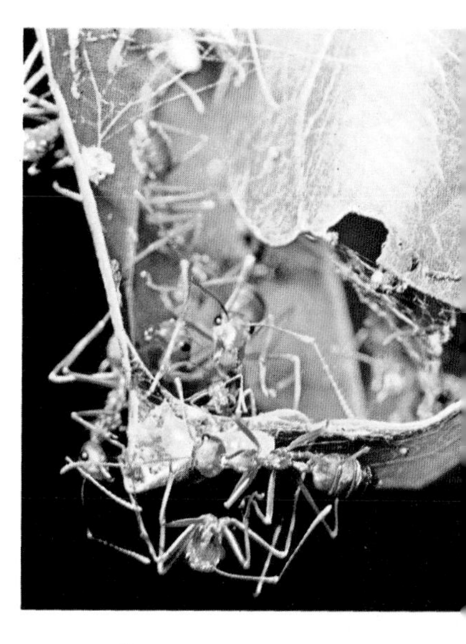

***Oecophylla* ants, a common jungle type, construct nests by sewing together the edges of large leaves. At the bottom, one ant holds a plump white larva, which spins the threads of silk that bind the nest into a secure shelter.**

Many of the insects remain hidden from sight, but others may be painfully evident. If you stand still for a moment, you are almost certain to be attacked by mosquitoes. And ants are likely to begin crawling up your legs, for the jungle seems to be alive with them—leaf-cutting ants, army ants, ants of all sizes and shapes. Some have painful bites or stings, but others are quite harmless.

Look around and you will soon discover ant's nests of many sorts. Some are on the ground or on rotten logs, while others are high in the treetops or even hidden inside the hollow stems of certain plants. Although it is difficult to determine how many ants live on an acre or even a few square yards of jungle, at least it is possible to count their nests. One zoologist tried to do this in the jungles of West Africa. He estimated that on every four square yards there was one nest of *Oecophylla* ants, which build nests of leaves sewn together with silken threads, and that one tree in five was inhabited by a colony of *Polyrachis* ants. In addition to these larger kinds, there were more nests than he could count of the smaller species. Every tree, in fact, had at least one nest of some kind of ant. When you consider that each nest harbors a colony of hundreds or even thousands of individual ants, you begin to realize just how many ants there are in the jungle.

The most formidable of all tropical ants are the army ants of South America and the driver ants of African jungles. These notorious insects do not make permanent nests but roam the forest in groups that may include a million or more individuals and sometimes cover hundreds of square feet. As the army progresses, the ants devour any insects and other small creatures they encounter. The African driver ants are especially dangerous; even pigs or fowl unable to escape their advance may be completely stripped of flesh.

Perhaps just as numerous as the ants are termites, or "white ants" as they are often called, though they are not really closely related to ants at all. Their strange-looking nests, made of soil particles or chewed wood, resemble fantastic towers or overgrown mushrooms, or form great bulges on tree trunks. The nests are a common sight in every tropical jungle, but the insects themselves generally keep well under cover. They live in galleries in rotten wood or in the soil and seldom appear in the open during the day. But if you slash at a rotten tree with your machete or pull the bark off of a dead log, you will see them swarming by the thousands.

Both ants and termites play an important role in the economy of the jungle, for they consume enormous quantities of plant and animal material, living as well as dead. And they in turn are eaten by birds, lizards, toads, and many other animals.

**The large globular mass high in a tree in Puerto Rico is an aerial termite nest, one of many different types of termite nests found in tropical rain forests. Termites, which feed chiefly on wood and other dead plant material, are some of the jungle's most active scavengers.**

Termites that live in aerial nests travel to the ground through covered passageways plastered to the tree trunk (*left*). The tunnel, composed of a mortarlike mixture of digested humus and soil, constantly teems with two-way traffic. Above, a portion of the tunnel has been broken away to reveal the soft-bodied inhabitants of the colony. Accustomed to living in total darkness, the industrious termites will hasten to repair the damage to their tunnel.

**Stout, curved claws enable the tamandua, or three-toed anteater, to rip easily through the walls of a termite nest. Seemingly just along for the ride, the half-grown young waits patiently as the mother feasts on a meal of plump termites.**

## *Animals that eat ants*

Some animals, in fact, specialize in a diet of ants and termites. Best known are the anteaters. They use their stout claws to rip open the insects' nests, then lap up the insects with their long, sticky tongues, which can extend well beyond the tips of their strangely elongated snouts.

The species most commonly seen in zoos is the giant anteater. This comical-looking, six-foot-long beast with an enormous, bushy tail lives on the ground in grasslands and jungles of Central and South America. In the tropical American jungles are two other species as well, the squirrel-sized two-toed anteater and the slightly larger three-toed anteater, or tamandua. Both of them live mainly in trees and are seldom seen. But if you are lucky, you may spot a tamandua using its thick *prehensile,* or grasping, tail like a fifth leg as it clambers from branch to branch.

Even more bizarre are the pangolins of Asia and Africa.

As with the anteaters, some species are *terrestrial*, or ground-dwelling, while others are *arboreal*, or tree-dwelling. Like the anteaters, they have stout claws on their forefeet for ripping open ant and termite nests and long sticky tongues for catching the insects. But their most distinctive feature is their covering of scales. From the tips of their elongated snouts to the ends of their long tapered tails, their bodies are almost completely covered by stiff, sharp-edged scales that overlap like the shingles on a roof. In times of danger, they simply roll up into tight balls, exposing only their armor of scales to their enemies.

The rather closely related armadillos also are covered with armor, but instead of overlapping scales, it is formed of hard bony plates. Like pangolins, some of the armadillos can roll up into balls in self-defense. They are common in the South American jungle, where they forage on the ground for ants but feed mainly on other kinds of insects.

Many jungle birds also feed on ants, but even more interesting than the species that actually eat ants are the ant-thrushes and ant-wrens. In American and African jungles, these birds follow roving bands of army ants and driver ants

**The silky or two-toed anteater, only fifteen inches long, is the smallest of America's three species of anteaters. Like both of its larger relatives, it uses its long sticky tongue to lap up its usual fare of ants and termites.**

and actually announce the ants' presence by their constant flitting and chirping. Sometimes they alight at the army's advancing front, but they do not settle long enough to be attacked. And instead of eating the ants, they live mainly on cockroaches, spiders, and other small animals that are disturbed by the marching ants.

**Like the boa, the little green python of New Guinea jungles is a nonvenomous constrictor. The row of pits on its lower jaw are sensory organs that help the python locate its prey.**

### *Some jungle snakes*

Some of the strangest jungle creatures are snakes. Many different kinds live in the rain forests of the world. But, contrary to popular notions, they do not dangle from every branch or lurk beneath every bush, and many of them are quite small. The visitor, in fact, is quite likely to go for days or even weeks on end and never see a single one.

In South America, some of the most dreaded snakes of all are the bushmasters, ten-foot-long relatives of our well-known rattlesnakes. They are the second largest poisonous snakes in the world, exceeded only by the king cobras of southeast Asia, which may reach a length of eighteen feet. Yet they are hardly common, and when they lie coiled on the jungle floor, their mottled patterns of browns and tans blend so well with the dead leaves that you can pass within a few feet of one and not notice it. At the approach of a man, they usually slither slowly from sight, and they are not likely to attack unless provoked.

The same is true of the boa constrictors, whose fearsome reputation is highly exaggerated. It is true that these heavy-bodied, twelve-foot-long snakes can kill sizable prey by coiling around their victims and squeezing them to death. But they are harmless to man and in fact make rather docile pets.

The largest of all living snakes, the giant anacondas, also live in northern South America. These near relatives of boa constrictors are said to reach a length of thirty feet, although twenty-five feet or less is probably closer to the truth. Giant anacondas lurk beside streams or pools, waiting for animals to come to drink. Excellent swimmers, they drag their victims underwater, crush them, and then swallow them whole.

**Perhaps the best known of all jungle snakes is the boa constrictor, a native of Central and South America. Despite its twelve-foot length and heavy body, the boa is an agile tree climber.**

## *Animals everywhere*

Near water is a good place to look for animals of other sorts. In the soft mud by streams are the footprints of many creatures that come for water. Here you may find the tracks of peccaries, piglike animals that wander through the jungle, sometimes in large herds. One kind, the collared peccary, ranges all the way from Paraguay to southern Texas, surviving as well in near-desert conditions as it does in jungles.

In South America you may also find the delicate tracks of brockets, tiny, short-antlered deer that stand only two feet high at the shoulder. Even smaller are the chevrotains, or mouse-deer, of tropical Asia and Africa. Although they lack antlers, these little hoofed animals are closely related to true deer. But even the largest stand only a foot high.

The South American jungle also is the home of some true giants. *Rodents*, for instance, are a group of mammals characterized by long, continuously growing front teeth that are

**Capybaras, the largest rodents in the world, grow to four feet long and weigh up to 120 pounds. These shy, gregarious animals, common along jungle streams and rivers, are frequently hunted for their hides, which are sold as "wild" or "natural" pigskin.**

especially suited for gnawing. In the jungle as elsewhere, the commonest, most familiar rodents are forms such as mice and squirrels, all quite small. In North America, the largest rodent is the beaver, which rarely weighs more than sixty pounds, and is about three and one-half feet long, including its tail.

But living in the rivers and lakes of eastern South America are animals called capybaras which are the largest rodents of all. They measure four feet long and weigh as much as 120 pounds. These astonishing creatures, which look something like overgrown guinea pigs, are expert swimmers. They feed on plants both in the water and along the shore.

Two other large rodents, both about the size of rabbits, also live in the South American jungles. Agoutis, about twenty inches long, run rapidly with a bounding gait but also can swim. The slightly larger sooty pacas are quite similar in appearance and habits. But both are active mostly at night and remain in hiding during the day.

**Three of the world's four species of tapirs live in the Americas, and the fourth is found in southeast Asia. Like all its relatives, this South American species has a drooping, prehensile snout that aids it in browsing on marsh plants.**

If you are lucky, you might startle a tapir hiding in a dense thicket near the river and hear it scuttling through the underbrush. Six feet long and three feet high, these long-snouted vegetarians look like some strange mixture of pig, cow, and elephant.

Many other animals spend part of their lives on or near ground level—iguanas and other lizards, small quail-like birds called tinamous, and many more. There are the fabled jaguars and their smaller relatives, ocelots and jaguarundis. But you are no more likely to see one of these than you are to see a bear or a mountain lion in a North American forest. They hunt mostly at night and by day usually slink away before they can be seen.

The same is true of many of the smaller animals of the jungle, such as the opossums, the West African bush babies, and the strange little saucer-eyed tarsiers of Borneo. These creatures are active only at night. You are likely to see one only if you happen to disturb it as it sleeps in a tree during the day.

### *A hothouse climate*

So far in our exploration of the jungle, we have been concentrating on the plants and animals of the rain forest. But what of the climate and the ways in which it influences their lives?

One thing you are sure to notice is the fact that the air is always quite warm. This is hardly unexpected, since all the great jungles of the world lie in the tropical zone, which is to say, within 23½ degrees north and south of the equator. But exactly how warm does it get in the jungle? Relatively few long-term weather records are available for tropical countries, but at Mazaruni Station in Guyana, just 6 degrees 50 minutes north of the equator, records have been kept for many years. Some of the statistics are especially interesting. For one thing, they show that the highest temperature ever reached is about 93 degrees—less than the

On the next two pages, a scene in a South American rain forest contrasts animal activity by day and by night. During the day, curassows, tinamous, and hummingbirds are active. Butterflies flutter through the undergrowth and anoles skitter up tree trunks. Larger animals such as brocket deer, coatimundis, and agoutis also are up and about and, occasionally, howler monkeys venture low enough in the forest canopy to be seen from the ground.

At night a whole new army of animals takes over. Silky anteaters, woolly opossums, kinkajous, and armadillos emerge from hiding to search for food. Fireflies brighten the darkness with pinpoints of light. Bats wing silently through the trees and, near clearings, tropical screech owls and nightjars swoop near ground level. And, as the frogs begin their nighttime chorus, jaguarundis and other large predators set out on their nocturnal hunts.

→

HOWLER MONKEY
BROCKET DEER
COATIMUNDI
GREAT CURASSOW
VARIEGATED TINAMOU
HERMIT HUMMINGBIRD
HELICONIUS BUTTERFLY
AGOUTI

SILKY ANTEATER
WOOLLY OPOSSUM
SPECTACLED OWL
KINKAJOU
TROPICAL SCREECH OWL
NINE-BANDED ARMADILLO
JAGUARUNDI
RUFOUS NIGHTJAR

high summer temperatures frequently recorded in many places in the United States!

On the other hand, however, temperatures in Guyana never drop very low. The records show that the average temperature for the year is 79 degrees. Even more significant is the fact that the average temperature for the hottest month is 81 degrees, while the average for the coldest month is 77 degrees. Indeed, the temperature difference between night and day may be 8 degrees or more—much greater than the difference in the monthly averages.

The jungle, then, is a hot but remarkably equable environment. Conditions, in fact, vary even less than these figures indicate, for the records at Mazaruni Station were taken in

a clearing. Inside the jungle in the shade of trees, the temperature range is even less. There the conditions are more constant than they are in many thermostatically controlled living rooms.

Another thing that surprises many newcomers to the jungle is the comparatively small amount of bright sunshine. At Mazaruni Station, for instance, the day is always about twelve hours long. Yet the sun shines on the average for only about five and one-half hours per day. The reason for this, of course, is the great amount of rainy and cloudy weather. The sky is seldom clear of clouds, and rain falls frequently, sometimes in violent thunderstorms but more often in heavy showers that end as unexpectedly as they begin.

**Although dry spells in the tropical rain forest occasionally last for two or three weeks, normally it rains on two days out of every three. Over the course of a year, rainfall usually totals one hundred inches or more, and in some jungles the annual total may exceed two hundred inches. Here, a veil of rain roils the surface of a jungle river in Surinam. Frequently, as on the next two pages, rainstorms are accompanied by spectacular displays of lightning.**

At Mazaruni Station the average annual rainfall is ninety-seven inches, more than eight feet! This is much greater than most places even in the wettest parts of the United States or Europe. In New York City, for instance, the average annual rainfall is about forty-two inches, while in Chicago it is about thirty-three inches and in San Francisco the total is slightly less than twenty-one inches.

Nor is the rain in Guyana restricted to any particular time of year. At Mazaruni Station it rains on the average on 212 days each year—nearly two days out of three. Some months are wetter than others, however. The highest rainfalls are in December, January, May, June, and July. But there is no really dry season. Although rainless periods of as long as three weeks have been recorded, no month has an average rainfall of less than four inches.

**Land crabs, like this one from Panama, are common in the constantly damp environment of the rain forest. They lurk in burrows and under rocks, but come out regularly to feed on rotting vegetation.**

## *Life in an ever-wet world*

From all this it is clear that "tropical rain forest" is a good name for the areas we have been calling jungle. As might be expected, the constantly warm temperatures and frequent rainfalls have profound effects on the life of the jungle. For

one thing, the rainfall often is very violent. Although the effects of heavy rain have not been studied very much, it seems quite likely that the battering of heavy raindrops could be a serious hazard to delicate leaves or small insects in exposed positions. Possibly the prevalence of tough, leathery, undivided leaves among rain-forest trees is in part an adaptation protecting them from damage by rain.

The constant humidity of the air is another obvious feature of the climate. In the interior of the rain forest, especially near the ground, the air is almost always so nearly saturated with moisture that very little evaporation is possible. You will notice this if you camp under the trees without first making a clearing. Unless they are exposed to direct sunlight, wet clothes never dry out and soon become moldy.

For many small plants and animals, this permanently moist atmosphere is very important. Because of it many filmy ferns with paper-thin leaves are able to flourish in the jungle. If the trees were cleared away, the ferns would quickly dry out and die from loss of water. The jungle also harbors many animals belonging to groups that in other climates live only in or near water. Thus land crabs are quite common near the sea in the South American jungle, and under bark and in rotten wood there are flatworms and even

**Snails also flourish in the humid jungle habitat. Some species live on the forest floor, while others live high up in the trees. Here a pair of Puerto Rican snails are mating, with one individual nearly hidden beneath the other.**

polychaete worms, a type that elsewhere is found only in water.

Leeches also usually live in water, but in the wet jungles of Southeast Asia some of the most notorious pests are land leeches. In Borneo or the Philippines, you would be lucky to return from a jungle trip without finding half a dozen of these unpleasant creatures on your legs or arms. It is remarkable how quickly they can attach themselves to passing men or animals. They pierce the skin with their sharp horny jaws, and drink blood until they are swollen to several times their original size. Only a lighted cigarette or a pinch of salt will make them loosen their hold.

A far more attractive group of jungle animals that depend on water or a moist atmosphere are the frogs and toads. Jungles harbor enormous numbers of species of these am-

phibians, ranging in size from tree frogs no bigger than a fingernail to bullfrogs whose tadpoles are nine inches long.

At least half of the jungle frogs live in trees, and many have large suction disks on their toes which enable them to climb on branches. Some of them, moreover, have developed means for producing young without laying their eggs in ponds and streams as most frogs do. In tropical America, some frogs lay their eggs in the water stored in epiphytic bromeliads. Other tree frogs lay their eggs in sacs made of leaves on branches overhanging streams. When they hatch, the tadpoles fall into the water. Still other species carry their eggs on their backs or deposit them in damp crevices. In such cases the tadpole stage is completed before hatching, and what eventually emerges from the egg is a tiny but fully formed frog.

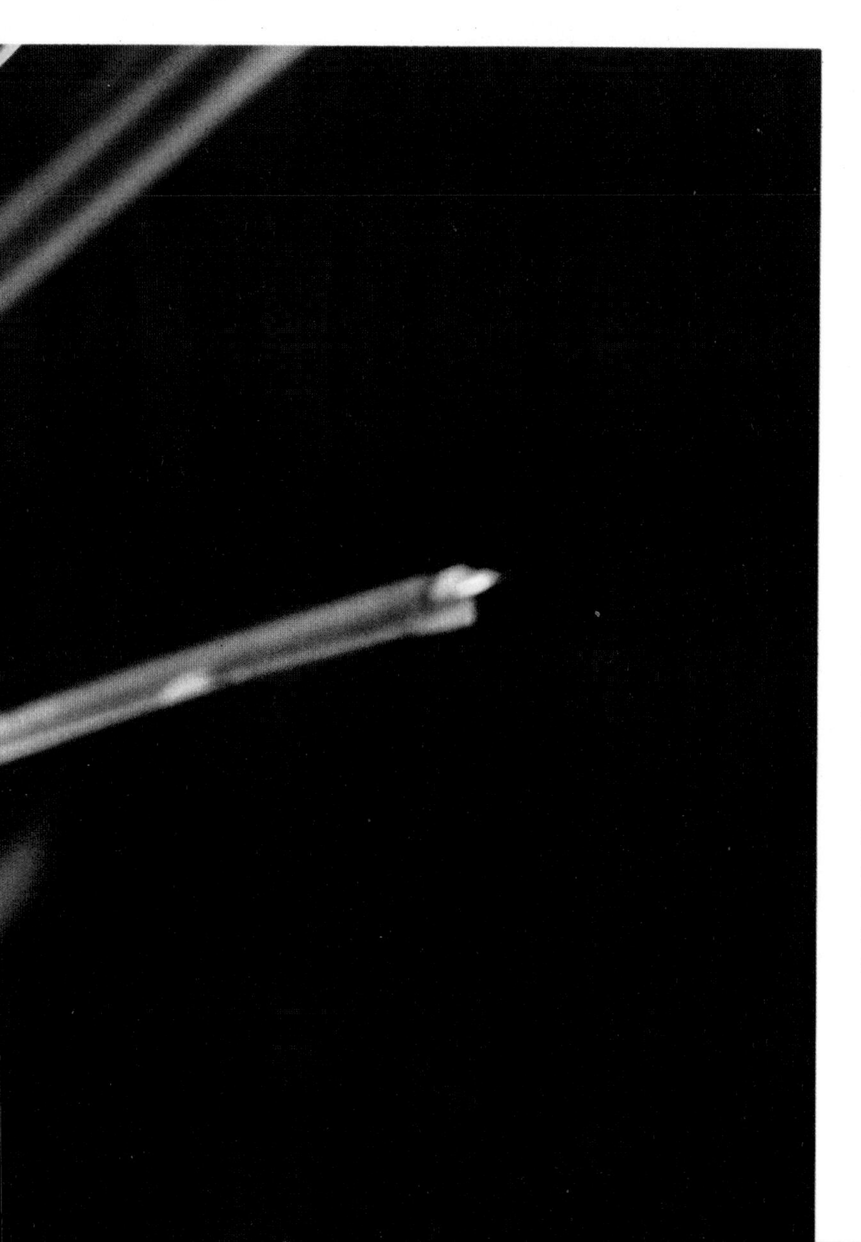

**Rain forests throughout the world are inhabited by many kinds of tree frogs, some of them as large as bull frogs, others as tiny as this one and one-half inch long *Philautus* frog from Asia. Yet frequently they go unnoticed—until they begin their loud nighttime choruses of croaks, trills, and whistles.**

In addition to the suction disks at the tips of their toes, tree frogs, like this Central American red-eyed tree frog (*left*), are distinguished by their long slender legs. Unlike most frogs and toads, which progress by hopping, tree frogs move by walking. Hence their legs are less muscular than those of their jumping relatives.

Among the most brilliantly colored—and notorious—of all frogs are the *Dendrobates* or poison-arrow frogs of South American jungles. Two fairly common representatives of the group are ***Dendrobates pumilis*** (*opposite page*) and ***Dendrobates aliratus*** (*above*). Long before the arrival of Europeans, Indians used these frogs as a source of poison for their hunting arrows. They obtain the poison, which is secreted by glands in the animals' skin, by killing the frogs with sticks and then holding them over a fire. The heat causes droplets of poison to ooze from the skin. Arrows dipped in this material are so potent that they can paralyze animals as large as monkeys. Even touching the frogs' skin can cause severe pain.

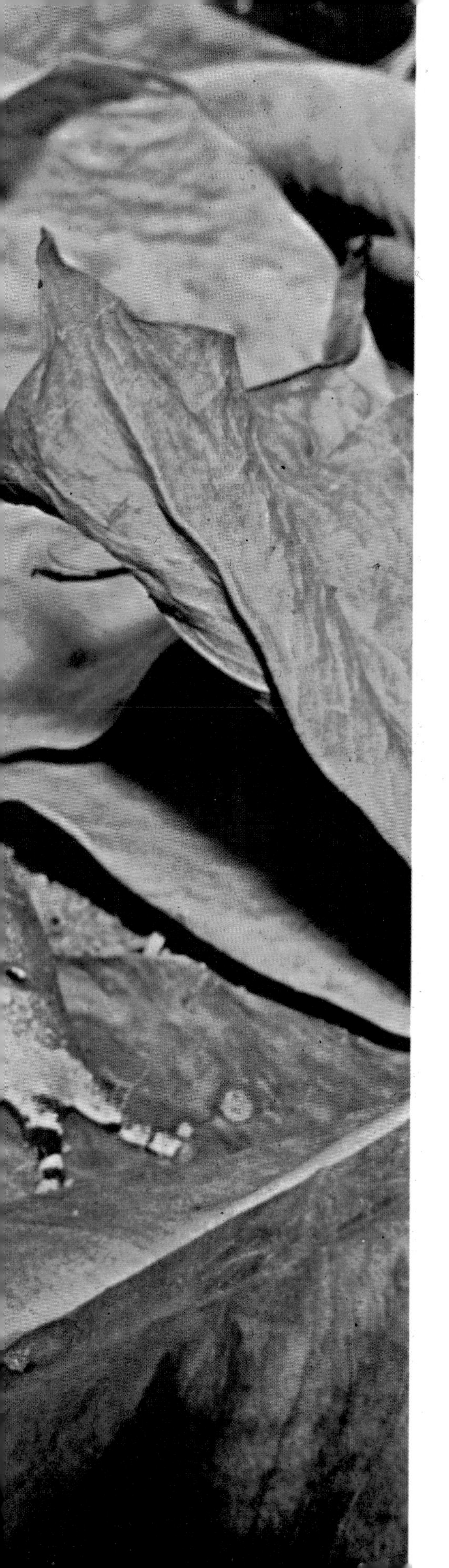

One of the most remarkably camouflaged of all jungle frogs is the Asiatic horned frog (*left*), a native of Malaya and Indonesia. Earthy colors and projecting hoods of skin over its eyes enable it to merge almost completely into the cover of dead leaves on the jungle floor.

The tiny tree frog from southeast Asia (*below*) is conspicuous against a background of green leaves. But when it rests on branches or among dead leaves, its brown colors and angular form seem to melt into its surroundings.

## *A world with no winter*

The jungle is a world with no seasons. Throughout the year, the climate remains unchanged, with frequent rainfalls and constant high temperatures and humidity. Whether it is July or January, plants here and there are always in bloom, leaves are always green, and the luxuriant growth of trees, vines, shrubs, and herbs continues without interruption . . .

Even more important than the warmth and humidity of the jungle is the fact that this hothouse climate remains more or less constant throughout the year. Because of the continually high temperatures, the life processes of plants proceed more rapidly than in cold climates. As a result, great quantities of the plant material upon which all animal life depends are constantly being produced.

The rain forest is a place where winter never comes; in the jungle it is always summer. Although some trees may shed their leaves all at once and remain bare for a few days or weeks at a time, there are always leaves on most of the trees. There is no time of year when plants cannot grow, flower, and produce fruit; there is no season when seeds cannot germinate.

JULY

JANUARY

In temperate forests, most of the trees are *deciduous.* They lose all their leaves in the fall, with the next season's leaves growing from buds that are fully formed by the end of summer. These winter buds, usually enclosed in protective scales, are well adapted for enduring the severe cold of winter.

In tropical jungles, on the other hand, most of the trees are *evergreen.* Like laurels and hollies, they are covered with green leaves throughout the year. Their leaves of course do not live forever. Although there is little precise information on the life spans of individual leaves of tropical trees, most of them probably live no longer than two years. But the leaves do not all fall simultaneously. New leaves are growing even while old ones are falling, so that the tree often bears old and young leaves at the same time. The shedding of old leaves and production of young ones may not even be simul-

. . . In temperate regions, in contrast, the differences between summer and winter are dramatic. In forests in the northeastern United States, for instance, all the trees are covered with leaves in summer, and plants bloom on the forest floor. But in winter, only a few evergreen trees retain living leaves, and life comes almost to a standstill until the return of warm weather in the spring.

taneous in different parts of the same tree. One branch or part of the crown is often at a different stage from all others on the tree.

Although the buds of jungle trees may be damaged by insects or eaten by squirrels or monkeys, they are not liable to damage by cold or drying winds. Probably as a consequence, jungle trees commonly produce fewer but larger buds than do trees in temperate zones. The palms are an extreme example of this phenomenon. They produce only a single large bud at the tip of each stem. If this bud is killed, the whole stem will die, since there are no other buds to take its place. And, in fact, in Brazil many palms are killed for the sake of their large buds known as palm cabbage, which in some species are very good to eat.

**In the tropical rain forest, caterpillars abound everywhere and come in an astonishing variety of shapes and colors, such as this bristly individual from South America . . .**

Like the trees, the small soft-stemmed or *herbaceous* plants of the jungle also lack adaptations to winter cold. In the United States and other temperate countries, many plants of the forest floor die down in late summer. In some, such as mayflowers and wild geraniums, the stems die down to ground level in the fall, with new growth in the spring coming from buds that remain hidden beneath the soil surface. In others, such as trout lilies, only a bulb, which is often buried deep in the soil, stays alive through the winter.

The herbaceous plants of the jungle, in contrast, do not die down periodically, and their buds are usually exposed well above ground level. And very few produce underground bulbs. Those that do are found mostly in jungles that have a yearly dry season rather than in the ever-wet environment of the tropical rain forest.

## *Easy living throughout the year*

Just as jungle plants are able to grow throughout the year in the rain forest, there is no season when animals cannot breed and feed. In cold climates, winter brings not only the risk of freezing to death but also the possibility of starvation. With their usual sources of food cut off, some animals such as frogs, lizards, snakes, and certain mammals survive the winter only by *hibernation*. Bats and woodchucks, for instance, remain in a torpid, sleeplike state for months on end. Their body temperatures drop nearly to that of the surrounding air, and their life processes slow to a minimum. Other mammals, such as certain squirrels, do not hibernate completely. They remain in nests and burrows where they are to some extent protected from the weather, and wake up periodically to feed on supplies of food they stored in the fall.

Jungle animals, in contrast, neither hibernate nor hoard food. Instead they can remain active throughout the year, for their food, though it may be more plentiful in some seasons than in others, is always available. And although most species breed at certain times of year and not at others, there is no season when some animals are not mating or bringing up their young.

Birds of northern climates also differ in a similar way from their tropical relatives. Perhaps because of their high body temperatures, most birds do not hibernate. Instead, when winter comes, many of them remain in their home areas, depending on their insulating feathers to keep them warm as they move about finding whatever food and shelter may be available. And a great many more, as we all know, migrate to warmer climates, often to the tropics.

Jungle birds, on the other hand, are mostly permanent residents. Since they have a year-round food supply and perpetually warm weather, they do not migrate from one country to another. At most they move over relatively short distances, possibly in search of favorite foods that at a particular time may happen to be more plentiful in one place than in another. Tropical birds that live outside the jungle in open areas such as in clearings, on savannas, or along riverbanks, however, sometimes migrate over quite long distances, especially in Africa. But their movements probably are triggered by food shortages in the dry season rather than by changes in temperature or day length.

**. . . and this Asian caterpillar. Since they live in a world without seasons, different species of jungle caterpillars are likely to be seen during every month of the year.**

Jungle insects, like the birds and mammals, also live in ways that would not be possible if there were a cold season. Insects of northern climates, for instance, usually have a *diapause* in their life histories, a period of dormancy comparable to the hibernation of larger animals. This "resting" period may occur at any stage in the insect's life from egg to adulthood. In our familiar moths, the diapause occurs in the pupal stage, when the caterpillar undergoes its transformation to the flying adult. During this stage, which often occurs in winter, the pupa remains in its cocoon in the ground or some other sheltered location.

But unless they live in an area with a severe dry season, tropical insects often have no diapause at all. Their life histories continue without interruption throughout the year. Instead of producing only one or two generations each year, generation can follow generation. For instance, the eggs, caterpillars, and adults of a swallowtail butterfly of the Philippines can be found all year round. And tropical wasps' nests, instead of being used for only a single summer as they are in northern climates, continue to be occupied year after year. When the nests become overpopulated, the wasps swarm and start new nests, just as our familiar honeybees do.

## *Trees for all seasons*

To some extent, the perpetual greenness and apparent lack of change in the jungle are an illusion, for closer examination reveals that many, perhaps most, individual species of trees do in fact show seasonal changes of some kind. Most species have their own times for flowering, fruiting, and producing bursts or flushes of young leaves. The seasonal changes of different species, and sometimes even of different individuals of the same species, do not all take place at the same time, however. This is in striking contrast to our own northern forests, where all the oaks, beeches, and maples come into bloom, produce fruits, and drop their leaves at more or less the same time.

Some interesting observations on leafing and flowering in tropical trees were made at Singapore, which has one of the most constant climates in the world. The highest temperature ever recorded there is about 93 degrees, the lowest, 70 degrees. Usually the temperature remains between 74 degrees and 89 degrees, and the average temperature for the hottest month is only 3 degrees higher than for the coolest.

At Singapore, some trees, such as the nutmeg, seem to grow and produce new leaves continuously, but the majority do not. The evergreens seem to produce their leaves in flushes at intervals that vary from species to species and from year to year. In some cases the flush seems to be triggered by a slight change in the weather, such as wet weather following an unusually dry spell.

**In the Ecuadorean jungle, a single blossoming tree provides a splash of color in a sea of green. Unlike temperate forests, where most trees bloom in the spring, in jungles different species bloom throughout the year.**

**Every January, as regularly as clockwork, the red silk cotton tree of Asian jungles bursts into bloom. The flamboyant display of crimson flowers on its leafless branches attracts many kinds of birds and insects that come to feast on pollen and nectar.**

The deciduous trees, on the other hand, seem to be somewhat more regular in behavior. In fact, leaf fall, followed by the expansion of young leaves, in some species has been observed to occur at regular yearly intervals. In other species, leaf change takes place every six months, while some species produce new leaves every seven to eight months and still others change their leaves at regular intervals of nine to ten months. In these cases it is difficult to believe that changes in weather can provide the necessary stimulus. It seems more likely that the trees have some kind of internal "clock" which regulates their behavior.

Flowering and fruiting seasons, of course, are even more important, since so many animals depend on flowers and fruits for their food. And, like the growth of new leaves, the production of flowers and fruits also seems to follow well-established rhythms.

In southeast Asia, some trees are ever-flowering. They begin to produce flowers and fruits when they are very young—sometimes no more than two or three years old—and remain continuously in flower as long as they live, which may be one hundred years or more. But these trees are definitely exceptions. Most species flower only at intervals such as once every six months, once a year, or even less often. Some of the merantis, which are among the most important tall jungle trees in Malaya, Borneo, and the Philippines, for example, blossom only once in every four or five years.

In some cases it is possible to find a distinct relationship between flowering and a particular kind of weather. Connections of this sort are most obvious in plants where a great many individuals flower at the same time. The pigeon orchid of Malaya, for instance, has long been known to bloom eight to ten days after a thunderstorm, with many plants coming into flower on the same day. Eventually it was discovered that the stimulus for flowering was not the storm itself but the sudden drop in temperature that accompanies a heavy rain. It has been found that the pigeon orchid's flower buds develop slowly up to a certain point but remain closed until they receive the necessary stimulus of a drop in temperature. In fact, a number of trees in the Malayan jungle have been nicknamed "temperature trees" because their behavior is similar to that of the pigeon orchid.

**Malaya's delicate epiphytic pigeon orchid is a jungle plant that blossoms in response to local weather conditions. Flowering is triggered by the sudden drop in temperature during severe thunderstorms.**

## *Animal seasons*

Just as different kinds of jungle trees have their own seasons for producing leaves, flowers, or fruits, different kinds of jungle animals also tend to have their own times of year for mating or rearing their young. Even though reproduction is feasible at any time of year in the unchanging climate of the jungle, the interesting fact is that most kinds of jungle animals do *not* breed throughout the year. Careful observation usually reveals that particular kinds of animals have rather well-defined breeding seasons. In the jungles of the New Hebrides in the Pacific, for instance, it has been found that various kinds of both insect-eating and fruit-eating bats have quite definite breeding seasons, even though the climate hardly changes.

Although relatively little is known about the breeding habits of most tropical mammals, birds have been more thoroughly studied in this respect. In the Guyana jungle, for example, where the two rainiest periods of the year are in May, June, and July, and again in December and January, the greatest number of bird species nest in May, that is, at the onset of the summer rainy season. A second less pronounced nesting peak occurs in September, just after the rainy months.

Some birds differ from the majority. Tinamous and curassows, which are fruit-eating ground birds, nest early in the year before the heavy rains begin, which is when they can find the most food. Hummingbirds, on the other hand, usually nest in the second half of the year, the season when the flowers on which they feed are most plentiful.

In nearby Trinidad, rainy periods and fruiting seasons are not quite the same as in Guyana. Here there is no month when some wild fruits are not available, and some fruit-eating birds are nesting at all times of year. Even so, nesting reaches a peak between April and June, when the greatest number of jungle trees are in fruit. The oilbird, which feeds its young entirely on the oily fruits of a few species of jungle

**With his bill hidden beneath his fanlike crest, a male cock-of-the-rock displays his elegant plumage. These spectacular South American cotingas concentrate their breeding season between February and April. During courtship the birds gather in clearings where the males perform elaborate displays before audiences of drably colored females.**

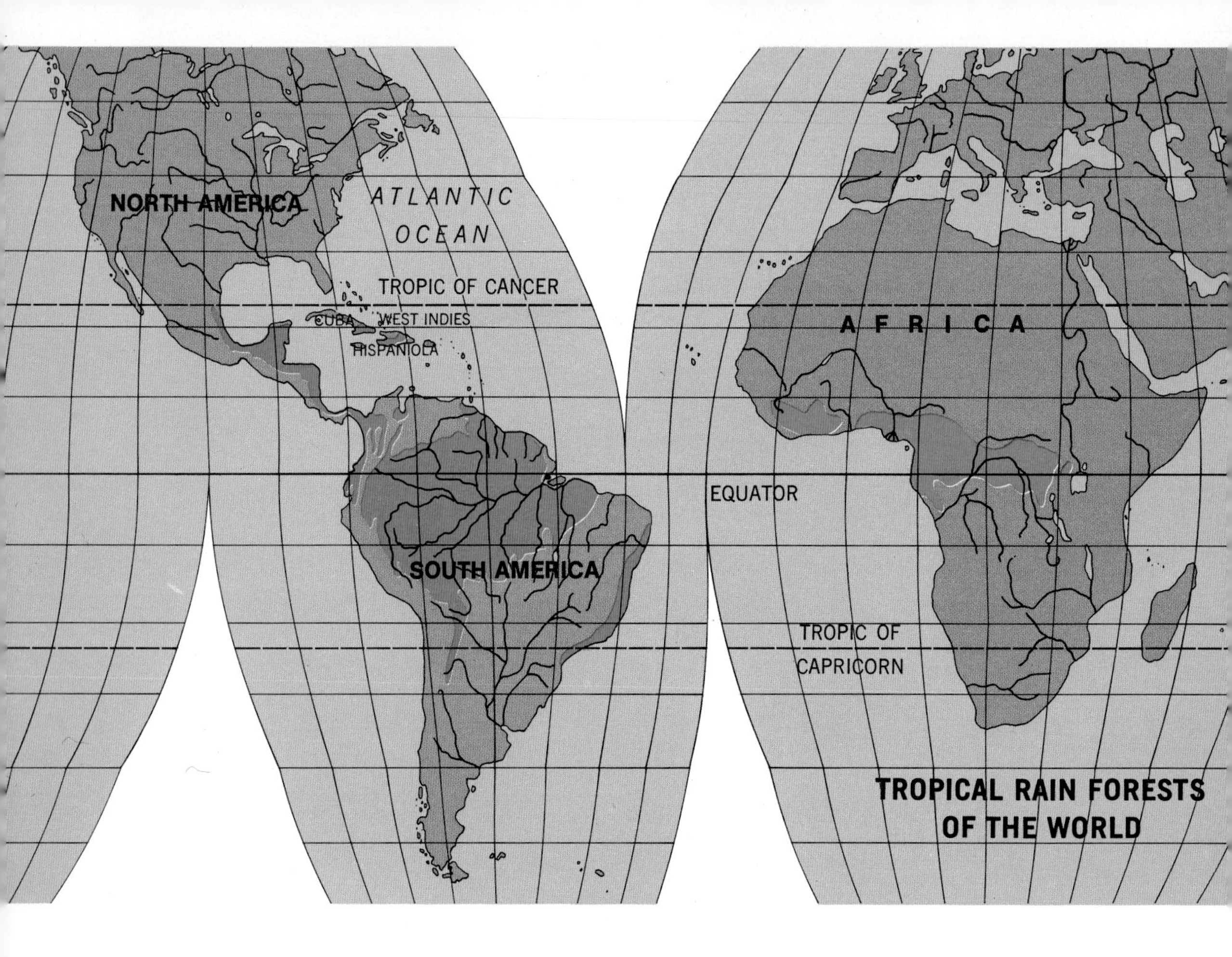


trees, mainly palms, shows this relation between nesting and the fruiting season particularly well.

Obviously there is a connection between the breeding season of many tropical birds and mammals and the time of year when their food is most plentiful. But this does not necessarily mean that abundance or scarcity of food is in itself the stimulus that controls their reproduction. In some tropical birds, activity of the reproductive organs is affected by very small changes in temperature or day length. Environmental changes of this kind undoubtedly play a role in determining their breeding seasons.

Yet other more subtle factors must play a role, too. For how else can we explain the fact that females of one species of bat in the New Hebrides of the southwestern Pacific Ocean become pregnant each year in the first few days of September—even though they spend all the daylight hours hanging upside down in caves where there is no light and temperatures scarcely vary?

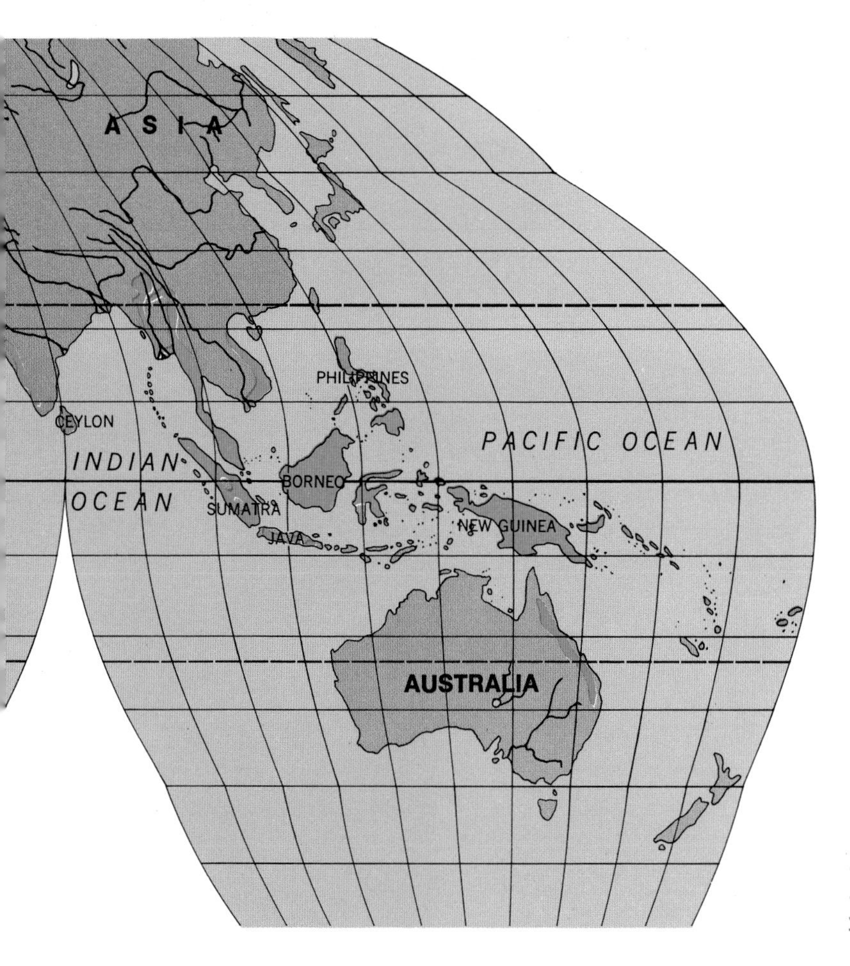


Tropical rain forests, shown in green, are widespread in tropical regions, but they flourish only under certain conditions. Although rain forests extend beyond the Tropics of Cancer and Capricorn in a few places, they grow best in the nonseasonal climate of equatorial lowlands, where temperature and humidity remain high and relatively constant and the rainfall is frequent and well distributed throughout the year.

## *Jungles around the world*

Rain forests are rapidly disappearing as trees are felled to make room for food crops and plantations of rubber, oil palms, and other crops. Yet in some places there are still thousands of square miles of tropical rain forests. Huge forests stand in South America, and there are others in Malaya, Borneo, the Philippines, Sumatra, New Guinea, and in Queensland along the northeastern coast of Australia. In Africa an immense jungle straddles the equator in the Congo, and a belt of jungle extends northwest along the coast of West Africa to the Cameroons.

The Old World rain forests are not as extensive as those of South America, but in many ways they are very much alike. Few of the plants and animals of the Old World jungles are identical to those of the New World, however. Indeed, some whole families found in one hemisphere are absent in the other. In moist areas of the American tropics, for instance, bromeliads grow everywhere as epiphytes and sometimes on the ground. But none are found in the Old World jungles, even where the climate seems quite suitable.

In the same way, the Old World has no sloths or armadillos, no hummingbirds, and no toucans. And although the Old World has plenty of monkeys, none of them have prehensile tails as do many New World species. Again, American jungles have no elephants, no apes, no rhinoceroses, no hornbills, and no plantain eaters, a group of birds peculiar to Africa.

On the other hand, there are many examples of unrelated families in one hemisphere replacing different families in the other hemisphere and playing almost identical roles in the life of the jungle. Thus while hummingbirds are a New World specialty, in the Old World there are flower-visiting sunbirds which pollinate blossoms in much the same way. The Old World hornbills and the New World toucans are not closely related either, yet they look much alike. Both birds have huge clumsy-looking beaks adapted for eating nuts and hard fruits. Tree frogs also are very numerous in both South America and tropical Africa, but the species found on each continent belong to quite different families.

Differences of this sort between Old and New World jungles are probably not due to differences in climate, however, but more likely to accidents of evolution and geologic events that took place in the distant past.

Very few kinds of plants and animals are found in both Old and New World jungles. Yet in both areas, many plants and animals of unrelated families have evolved similar adaptations for survival in the jungle habitat. Among birds, for instance, both the hornbills of Africa and Asia and the exclusively American toucans have grotesque bills especially suited for eating fruit and crushing nuts . . .

## *Different climates, different jungles*

In contrast to the ever-wet tropical rain forests we have been discussing so far, many tropical forests grow in climates with an annual dry season of two to four months. Great expanses of these *seasonal forests* are found in Central and South

America, in Africa, and in Asia, especially in India, Burma, Thailand, and Vietnam.

During the dry season no rain falls, or not enough to supply much water for plant growth. In some cases, the effect of lack of rain is increased by drying winds, as in West Africa where a dry wind known as the harmattan blows from the Sahara for several weeks at the height of the dry season. When this happens, herbaceous plants such as ferns wilt and dead leaves on the ground become so dry that they crackle underfoot.

The presence of a dry season does not always mean that the year-round climate is particularly dry. In a seasonal jungle, the yearly total of rainfall may equal or even exceed that of a rain forest. Cherrapunji in India, for instance, is said to have the highest rainfall in the world—an annual total of up to 460 inches, or nearly 13 yards! Yet it has a four-month-long dry season when more water is lost through evaporation than is received as rainfall or dew. As a result, the jungle is of the seasonal and not rain-forest type.

In many ways these dry-season jungles resemble the tropical rain forests. The trees are as big; in fact, it is here that some of the world's finest tropical timbers such as the mahoganies, teak, and the West African iroko are found. They have almost as many plant and animal species as rain forests. Indeed, many species are found in both types of jungle. There are just as many ants and termites, and both types of forest include a great variety of tree-dwelling animals. The abundance of large vines and buttressed trees are other features they share with rain forests, although epiphytes usually are less plentiful. During the wet months of the year, in fact, the general appearance and living conditions of the two types of forest are so similar that only an expert botanist or zoologist could tell them apart.

But in the dry season, the differences are quite pro-

**. . . Similarly, hummingbirds are found only in New World jungles. Yet their Old World counterparts, the sunbirds of tropical Asia and Africa, also have iridescent plumage, the ability to hover before flowers, and long slender bills adapted for probing for nectar and tiny insects hidden at the bases of long, tubular blossoms.**

HUMMINGBIRD

nounced. Although the seasonal forest as a whole never becomes as bare as an American forest in winter, during the dry season many trees lose their leaves. It is early in the dry season, moreover, that most of the trees come into blossom. And many birds and mammals tend to breed at the beginning or towards the end of the dry season.

In these jungles, plant and animal activity is not such a non-stop, all-year-round affair as it is in rain forests. The most active periods occur when there is a lot of sunshine and moisture. Some animals that depend on moist conditions, such as frogs, may even hide in hollow trees or disappear underground during the dry weather. In short, the seasons here are much more distinct than in the rain forest. But they are wet and dry seasons, not summer and winter.

In the tropics, especially near seacoasts, high mountain slopes are often covered with cloud forests. When warm, moisture-laden winds sweep in across the ocean and are deflected upward by the mountains, the air expands, grows cooler, and the moisture condenses, covering the mountaintops with mist and frequent rains. As a result, at higher and somewhat cooler altitudes, the tall jungle on lower slopes gives way to cloud forests, typified by dwarf trees, usually draped with luxuriant growths of mosses, lichens, and other epiphytes.

### *On mountain slopes*

Still another type of jungle is found in the tropics. If you climb the eastern slopes of the Andes from the Amazon Valley, or make your way up the great volcano, Cameroons Mountain, from the Gulf of Guinea in West Africa, or any of the high mountains of the Philippines, Borneo, or New Guinea, you will notice many changes along the way. The air becomes cooler at higher elevations and many of the lowland plants and animals disappear, their places being taken by other species adapted to lower temperatures. Above about three thousand feet, plants and animals related to those of temperate regions begin to appear.

Some of these mountain forests are among the wettest places on earth. As a result, the trees, which are much shorter and more crooked than those in lowland forests, are covered with fantastic blankets of mosses, lichens, and small ferns. These masses of epiphytes, in fact, are sometimes thicker than the tree trunks and branches.

The wet climate results because when moist air currents are deflected upward by the mountain slopes, the air grows cooler, causing the moisture to condense. As a result it rains almost daily and even when it is not raining, the landscape is often shrouded in mist. This is why these mountain forests of the tropics are called *cloud forests.*

A somewhat similar climate prevails in the northwestern United States on Washington's Olympic Peninsula. Damp

## CARIBBEAN NATIONAL FOREST

For North Americans, the most readily accessible of all tropical rain forests is the 12,500-acre Caribbean National Forest in northeastern Puerto Rico. The tract, administered by the U.S. Forest Service, is within easy driving distance of the capital city of San Juan.

The forest, located in the Luquillo range of mountains, centers on the island's highest point, a 3486-foot-high peak called El Yunque. Here, rising air currents drop an annual total of over two hundred inches of rain on the mountain slopes. Throughout the forest, dozens of rushing streams like this one tumble down rocky courses, shaded by the lush vegetation that flourishes in the constantly warm, moist climate.

*Like a lacy parasol, the fronds of a tree fern trace a delicate pattern against the sky in Caribbean National Forest. Since they require partial sunlight, these plants thrive best in clearings, along streams, and beside roads, especially in hilly country.*

*With his translucent throat pouch sagging like an enormous double chin, an agile anole pauses momentarily on a tree trunk. These little arboreal lizards, common in Puerto Rico and throughout tropical America, feed primarily on flies, spiders, and similar prey.*

*On the next two pages, a veil of mist softens the rugged contours of El Yunque, the focal point of Caribbean National Forest.*

*The alert visitor to Caribbean National Forest is almost certain to see hummingbirds flitting among the trees. This female Antillean Mango, tending an almost fully grown young, is one of the five kinds of hummingbirds found in Puerto Rico. In all, 319 species of hummingbirds—most of them tropical—are distributed across much of North and South America.*

winds blowing in from the Pacific drop an average annual rainfall of 142 inches in the western valleys of the Olympic Mountains, making this the wettest area in the continental United States. The resulting forest is indeed junglelike in its luxuriance. There are giant Douglas firs, western hemlocks, and other evergreens, some nearly three hundred feet high, and Oregon maples draped with mosses and liverworts. This forest is truly as spectacular as any South American jungle and part of it has been permanently preserved in Olympic National Park. But the climate is not tropical; the winter is quite cold, and there is a substantial snow fall. Because of this, the plant and animal life is not nearly so rich or varied as in a tropical forest.

All these other types of forest that we have been mentioning have some characteristics in common with the tropical rain forest. But each is unique; each could be the subject of a separate book. Here we will concentrate on the luxuriant evergreen forests of the tropical lowlands. These are the true tropical rain forests and are what most people mean by "jungle." They form a lush green world of towering trees decked with epiphytes and festooned with dangling lianas. And they are a world filled with mysteries well worth exploring.

**In the depths of the Malayan rain forest, a mud wasp completes the construction of her nest. Later she will stock the fragile structure with caterpillars and lay her eggs. After the eggs hatch, the wasp larvae will consume this food. Then they will transform into adult wasps, emerge from the nest, and renew the unending cycle of life in the mysterious world of the jungle.**

# *The Architecture of the Jungle*

When you stand on the floor of the rain forest, you see only a small fraction of the life of the jungle. Twenty or thirty feet overhead, the tree trunks disappear into a solid roof of foliage. Yet the tallest trees in the jungle are 150 to 200 feet high or even taller. The vast world of the treetops thus encompasses far more living space than there is on the jungle floor. And from the buzzing of insects, the whistling of birds, and the fruits and flowers dropped by troops of feeding monkeys, you know that the treetops must teem with unseen life. One naturalist, in fact, has commented that remaining on the ground in the jungle is "like standing on the floor of a great cathedral while the service is being celebrated on the roof."

Yet it is no easy matter to explore the hidden world of the forest *canopy,* that vast overhead realm formed by the branches and leaves of the tallest trees. You can learn little by looking up from the ground. The mass of foliage is too dense and even where an occasional gap permits a glimpse of the taller treetops, the trees themselves appear foreshortened.

Scientists have learned a great deal about the inhabitants

**An observation platform high in the trees provides a superb vista of the billowing, multicolored canopy of the Malayan jungle. Vantage points such as this provide scientists with unexcelled opportunities to observe the teeming life of the treetops at close range.**

of the treetops, however. Some of the knowledge has been gained by shooting individual animals as they moved among the branches. Rifles have also been used to shoot the tips off branches so that tree blossoms can be studied more easily. Some investigators have even used specially trained monkeys to collect small branches, along with their cargo of insects and other small creatures. And sometimes scientists cut down whole trees or even clear entire areas of trees in order to learn more about the makeup of the forest.

Of course, the best way to study the life of the canopy is to climb the trees and meet its inhabitants where they live. But to do this you must contend with many hazards, such as biting ants, thorny lianas, and rotten branches. Researchers have sometimes climbed the trees by using a special gun to fire a rope over a high branch and then climbing up on the rope. One even better, though more costly, method for exploring the life of the canopy is to build towers with observation platforms at various levels. Some students of the jungle have built platforms far above the ground where they could remain for days or even weeks at a time and observe the bustling life all around them.

By using all these methods, we have gradually accumulated a great deal of information about the architecture of the jungle and the life of the treetops. To the untrained observer, the mass of foliage overhead may appear to be a hopeless tangle of greenery interlaced by perplexing snarls of lianas. But even in this seeming chaos there is order.

## *The view from the air*

A good way to begin exploring the canopy is from a low-flying airplane. From high in the air the jungle looks like a vast blanket of green that billows across hills and valleys without a break. But when the plane sweeps lower, it becomes obvious that not all the trees are the same height. Here and there taller individuals project well above their neighbors so that the canopy appears hummocky and uneven. Sometimes two or three of these giant trees grow close together, but more often they are so widely separated that

**In tropical America, some of the gaudiest—and noisiest—inhabitants of the treetops are macaws, such as this scarlet macaw. Much in demand as specimens for zoos and frequently hunted for the sake of their feathers, macaws everywhere are dwindling in numbers.**

Some of the most important timber trees of India, Malaya, and the East Indies are various species of dipterocarps, giant trees that often grow 150 to 180 feet tall. Like other jungle trees, they usually occur in a mixture with many other species, but some dipterocarps are remarkable for growing in unmixed forests of a single species. Here, removal of neighboring trees has revealed the characteristic straight trunk and umbrella-shaped crown of a huge specimen of *Dipterocarpus bourdillonii*, a native of India.

the crowns of neighbors do not touch. From an airplane, they look like huge cauliflowers scattered through the forest canopy.

For the most part these tallest trees are about 110 to 150 feet high. In tropical America trees more than 150 feet high seem to be rare. In Malaya, on the other hand, jungle trees quite commonly exceed 200 feet, perhaps because both climate and soil are exceptionally favorable. Their crowns are usually very wide and heavy, supported by massive spreading branches so that the trees resemble gigantic umbrellas. In some types of jungle most of the tall trees are the same species, but usually they include many different kinds of trees.

Although these giants of the rain forest do not by any means form a continuous layer of foliage, they tower well above their neighbors. If the jungle can be likened to a house, they are in a sense its attic. In the language of botanists, the crowns of these tallest trees form what is called the A-story.

### *Life in the attic*

The tops of the A-story trees form a *habitat,* or living place, very different from the smaller trees or the forest floor. They are completely exposed to the burning tropical sun and, when it rains, they bear the full intensity of the storm.

Like everywhere else in the jungle, the most numerous animals in the A-story are insects. Some—the morpho butterflies—you might see even from a low-flying plane. Their large metallic-blue wings glinting in the sun are visible for long distances. Many species of morpho butterflies live in jungles but, except along rivers and in clearings, they seldom venture down near ground level.

Further evidence of the superabundant insect life is the swifts that zoom and dart above the trees, especially at dusk. Like whippoorwills, nighthawks, and other birds that capture insects on the wing, their bills, though short, are exceptionally broad enabling the birds literally to scoop their prey from the air as they maneuver gracefully on long narrow wings.

Now and then vultures circle lazily through the sky, constantly scanning gaps and clearings for the dead animals on which they feast. These great, broad-winged birds scarcely flap their wings at all but depend instead on rising air currents to keep them aloft.

The great harpy eagles of the South American jungle, in contrast, fly with rather heavy wing beats. But despite their great size—they are among the largest of all eagles—they maneuver quite well among the branches of the upper canopy on short rounded wings. The appearance of a harpy eagle winging silently through the trees can mean sudden death to a monkey or a sloth, their major sources of food.

The closely related Philippine monkey-eating eagle also specializes in monkeys, although it sometimes eats other prey as well. This bizarre-looking bird with a great feathered crest lives only in a few remote jungle areas in the Philippine Islands and is in grave danger of extinction. It is believed that no more than about fifty pairs of the birds remain in existence.

### *Down from the treetops*

If somehow you could be lowered from your plane into the denser vegetation below the scattered crowns of the A-story trees, you would find another distinctive layer of life in the

**An araçari's colorful bill immediately identifies him as a member of the toucan family. In New World jungles, several species of these birds dwell in the sunny intermediate levels of the rain-forest canopy. Like all toucans, araçaris tend to move about in groups and feed primarily on nuts and other fruits.**

jungle. This, the B-story, extends from about 110 to 30 feet or so above the ground.

Here there are masses of trees of many different species. But nearly all of them have straight slender trunks, usually not more than a foot in diameter, and rather small narrow crowns. If the A-story trees can be compared to umbrellas, those of the B-story look more like mops.

Usually, but not always, the trees are more closely spaced than those in the A-story. And here a great many more lianas twine among the branches, often binding one tree to another. But it is nothing like the dim world of the forest floor. The B-story is still quite a sunny place. Much of the space is free of branches and leaves, so that a good deal of light is able to penetrate to the even smaller trees down below, and there is plenty of room for birds and mammals to fly, glide, or jump about.

The interwoven vines and branches in the B-story are the home of many forms of life, especially birds. Many kinds of cotingas live here, as well as hawks, owls, pigeons, puffbirds, curassows, hummingbirds, and a host of others. Noisy, colorful parrots climb among the branches, sometimes using their beaks as additional limbs to assist in climbing. Parakeets, macaws, toucans, and trogons also provide frequent flashes of color.

**One of the handsomest of all parrots is Brazil's twelve-inch-long golden conure. This one is drinking from the "tank" of an epiphytic bromeliad, a very convenient source of water for a bird that spends most of its time high up in the trees.**

Many animals besides birds live in the treetops. Tree frogs are amazingly plentiful, and everywhere there are insects. Slender tree snakes slither among the leaves, and geckoes and other kinds of lizards dart along branches and on tree trunks. But even from a hide-out in the canopy you can rarely see these creatures. Most of them are camouflaged by their green or brown coloration and if you approach them, they either "freeze" or noiselessly dart away.

### *Slothful sloths and flying lizards*

Nor is there any shortage of larger animals. In South America you may find an astonishing creature nearly two feet long hanging upside down from a branch. This is a sloth, possibly the three-toed sloth or perhaps the slightly smaller two-toed sloth. These strange animals cling to the branches with their remarkably long, strongly curved claws. They often hang motionless for hours at a time, and when they change positions, it is usually with slow, lazy movements.

The sloths are easily overlooked. The long coarse hairs of their fur are grooved, and in the grooves live algae, microscopic green plants that give the animals' coats a curious greenish-grey tinge. This, combined with their habit of hanging motionless, makes them look more like ant's nests or masses of foliage than living animals. And the camouflage works. Eagles and jaguars, their major enemies, sometimes pass quite close to sloths without noticing them.

But for the most part the treetops belong to active, more agile creatures. The birds, of course, can fly. So can the many kinds of bats that live up in the trees, mostly sleeping by day and moving about at night to feed on insects, fruits, or pollen.

A number of other jungle animals also have become adapted for flight of one sort or another. The commonest are the flying squirrels, of which there are many species.

**Dangling from its daytime roost in the New Guinea jungle, a two-pound flying fox is nearly engulfed within its own stupendous wings, which may span as much as five and one-half feet. Far less fearsome than they look, these largest of all bats are strictly vegetarian, feeding entirely on fruit.**

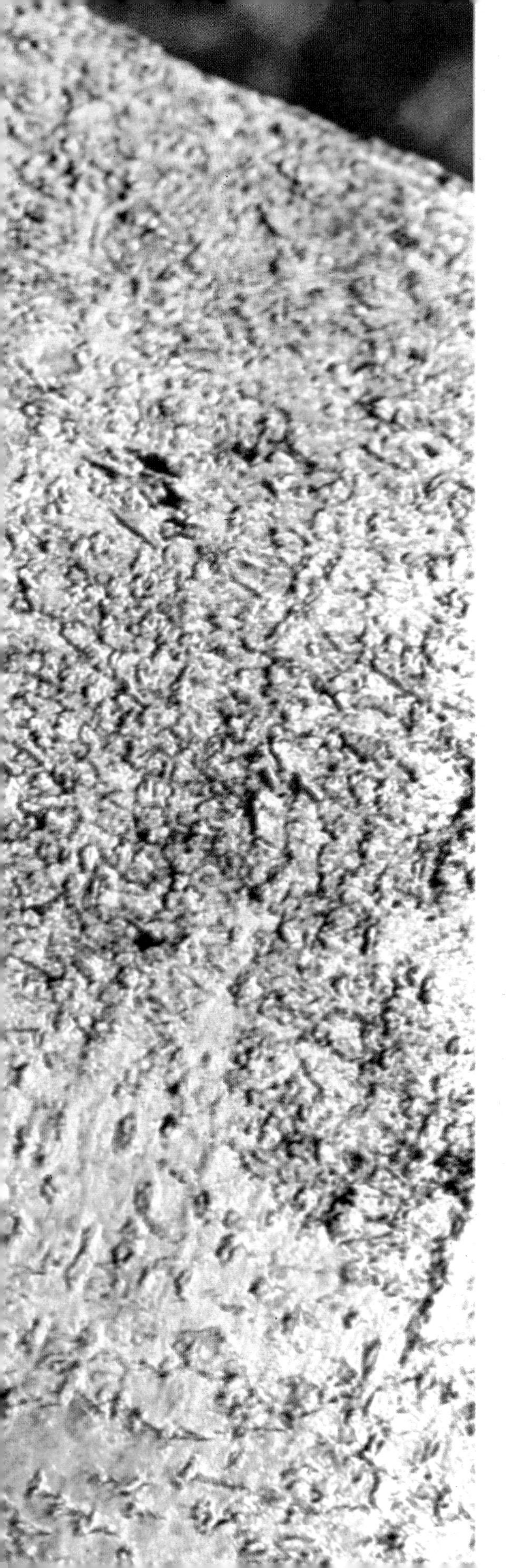

The upper levels of the rain forest, where the crowns of trees are fairly widely spaced, are very much the realm of creatures that can leap, glide, or fly from place to place. These animals have evolved an astonishing variety of adaptations that increase their mobility in the air.

When Wallace's flying frog of Malaya (*left*) wants to "fly," it simply leaps into space, spreads its broadly webbed toes, and arches its back to form a simple airfoil. Soaring gracefully downward, this bantamweight frog may travel forty feet or more before landing on the trunk of a nearby tree.

Southeast Asia's flying dragon (*below*), on the other hand, flies in an entirely different way. Projecting from each side of its body are several elongated movable ribs connected by membranes of skin. As it glides downward with these "wings" outspread, the ten-inch-long lizard looks something like a huge tropical butterfly. When the lizard runs along a branch, however, the "wings" fold back and seem to disappear along the sides of its body.

A view of its underside reveals the flying gecko's adaptations for two types of locomotion. Adhesive ridges on its toes enable the eight-inch-long lizard to climb on tree trunks and branches. For flight, it is equipped with webs between its toes and with flaps of skin along its legs, sides, cheeks, and tail. When the lizard is at rest, like this one photographed through a plate of glass, the skin flaps fold around its body. But when it launches into space, the flaps open like parachutes, permitting the gecko to make short, gliding leaps from tree to tree.

**The flying lemur, or colugo, of southeast Asian and East Indian jungles is a remarkable nocturnal glider, able to cover as much as two hundred feet in a single swoop. As it soars downward, broad flaps of skin extending from the sides of its neck to the tip of its tail make the colugo look like a large animated kite (*right*). By day, as a mother hangs slothlike from a branch, her flight membranes form a snug hammock to cradle her young (*below*).**

**Ability to dangle by its prehensile tail frees a howler monkey's forepaws for more useful purposes. Two feet long and weighing up to thirty pounds, howlers are the largest of American monkeys. They rove through the treetops in strictly organized tribes and feed on leaves and fruit.**

All are equipped with loose folds of skin between the legs on each side of the body. When they spread their legs, the squirrels are shaped something like kites and are able to glide long distances from tree to tree.

Much stranger is the colungo, or flying lemur, of southeastern Asia and nearby islands. This cat-sized animal has furry membranes extending from the neck to the forelegs, from the forelegs to the hindlegs, and from the hindlegs to the tip of the tail. When it launches itself into the air, the colugo is said to be able to glide two hundred feet or more in a graceful swoop.

In the southeast Asian jungles even lizards can soar through the air. The eight-inch-long flying gecko has membranes of skin almost all the way around its body that increase its surface area when it glides. Even better equipped for flight are the several species of flying lizards sometimes known as flying dragons. Along their sides, several ribs extend outside the body and are connected by membranes. Although they cannot be flapped, these fanlike structures truly resemble wings. When the lizards land, the supporting rods fold back along the sides of the body and the brightly colored "wings" disappear.

## *Monkeys and apes*

Other conspicuous and notably agile dwellers of the B-story are the monkeys which scamper among the branches and swing or leap from limb to limb. As we have already noted, many of the New World monkeys are further assisted in their climbing by the possession of prehensile tails.

Among the best known of the New World types are the howler monkeys, for although they are seldom seen from the ground, their calls are a common sound in the jungle. These large monkeys move through the treetops in groups and, regularly join in unearthly choruses, especially at sunrise and sunset. Their weird howls, amplified by special resonating chambers in their throats, are so loud that they can be heard a mile or two away.

Capuchin monkeys, marmosets, spider monkeys, and squirrel monkeys are other common New World types. The Old World, on the other hand, has such distinctive types as the langurs, colobus monkeys, macaques, and mangabeys.

Lanky, long-limbed spider monkeys of New World jungles are marvels of agility. These diminutive creatures fearlessly span gaps up to thirty-five feet wide in a single leap and can swing through the treetops as fast as a man can walk. Here a mother forms a living bridge to assist her timid baby across the open space between two trees.

Some of the great apes also spend much of their time in the trees. Two types found in southeast Asia are the orang-utan and gibbons, of which there are several kinds. Gibbons, which rarely grow more than three feet tall, are by far the more agile of the two. They move easily through the trees, swinging arm over arm as they hang from branches and even leap from branch to branch. In the Borneo jungle, their joyful bubbling cries are one of the most delightful early-morning sounds.

The orang-utan, in contrast, moves more cautiously, usually gripping supports with both hands and feet. When resting, it sleeps on a roughly built platform made of sticks and broken branches, thirty feet or more above the ground.

The two apes of Africa, the chimpanzee and the gorilla, are not quite so arboreal, or tree-dwelling. The highly intelligent chimps frequently move through the trees in family groups, foraging for fruits and other plant food. Like the orang-utan, they build platformlike nests of twigs and leaves in the forks of large trees. But they spend much of their time on the ground as well.

The gorilla, a much rarer animal than the chimpanzee, is simply too big and heavy to climb easily. A mature male gorilla may be six feet tall and weigh 450 pounds or more. In consequence, these largest of all the apes spend most of their time on the ground. Females and young sometimes climb trees, however, although they do so rather cautiously. And they often resort to the trees at night to build the nests where they sleep.

## *Life at lower levels*

Many other strange and wonderful animals live in the B-story. There are the shy, nocturnal galagos, or bush babies, and the closely related pottos in Africa. Here, too, one finds the scaly pangolin. Another strange beast of the B-story

**A young orang-utan, momentarily separated from its family group, clings precariously to a tree trunk. By the time it is full grown, this native of low-lying jungles in Borneo and Sumatra may weigh as much as 250 pounds. Because so many have been collected for zoos, these handsome animals are in grave danger of extinction.**

is the tree hyrax or African tree bear. Although its loud, croaking calls can be heard in the jungle almost every night, this woodchuck-sized animal is seldom seen. The tree hyrax looks ill-equipped for an arboreal existence, for its flattened toenails form little hooves something like those of a rhinoceros. But moist, naked pads that form suction cups on the undersides of its feet give it a firm footing on branches.

Yet very little of this life is visible from the forest floor. Below the B-story, at about twenty to thirty feet above the ground, is still another distinctive layer of vegetation. This layer, the C-story, consists of the tops of small trees with very thin, often polelike stems. It forms by far the densest mass of foliage in the rain forest and cuts off a large percentage of the light that manages to filter down through the A- and B-stories. Looking up from below, you can in most places see only the dense foliage of the C-story. Like a screen, it blocks out your view of the life in the upper stories.

Because it is so dense, the C-story hinders the movements of animals, such as birds and butterflies, from the upper stories. Yet it is not without life of its own. Birds such as wrens, manakins, and antbirds feed here as well as at lower levels. And many small animals such as bush babies and opossums live among the branches. Larger animals also make their way through the foliage on their way to and from the upper stories.

As we have seen earlier, the space beneath the C-story also is filled to some extent with vegetation. This layer of undergrowth, which generally is rather sparse, is called the D-story, and close to ground level is the E-story of small herbaceous plants and ferns. Some of the undergrowth consists of small palms. There are also woody unbranched "treelets," like trees in minature, that flower when they are no more than four or five feet high.

But most of the plants are seedlings and saplings of the

**Two elusive, rarely seen dwellers in the lower levels of jungle vegetation are the six species of African bush babies (*left*) and the three kinds of tarsiers (*right*), which are found on only a few islands in the East Indies. Large owlish eyes hint of the nocturnal habits of both of these agile, squirrel-sized relatives of monkeys.**

larger jungle trees, awaiting a gap in the canopy that will provide the light and space they need to shoot up into one of the upper stories. Only a few reach the higher levels, however. Most linger on for years as spindly saplings and die before they reach the flowering stage.

Cauliflory, the production of flowers and fruits directly on the main trunk and larger branches of trees and vines, is a common phenomenon in the jungle. It is found chiefly on smaller trees of the jungle, such as various species of cola trees.

Here in the lower levels we also find many trees and vines with the unusual habit of producing flowers and fruits on their main trunks and larger branches instead of in the usual position on smaller twigs. The flower buds of these *cauliflorous plants* push their way through the bark to the exterior. A well-known example is the cacao tree, a native of the Amazon jungle. In Guyana and in tropical Africa and Asia, many other small trees of various families also bear their flowers in this unusual way, though, curiously, very few cauliflorous plants are found outside the tropics.

The jungle, then, can be compared in a sense to a vast building with an attic, several stories, and a ground floor. It even has a basement, below the ground, where you can find a great deal besides roots. There are burrowing mammals such as armadillos and many kinds of rodents, as well as certain kinds of snakes and lizards that spend little or no time above the surface. Insects of many kinds, mollusks, and stupendous earthworms, some a yard long, also live in the soil. And of course there are innumerable fungi, bacteria, and protozoa so small that they cannot be seen without a microscope.

It is a rather shallow basement, however, since most jungles grow on sticky clay soils into which tree roots cannot penetrate very far. Nearly all the tremendous underground activity is concentrated in a layer no more than about three feet deep. Only where the soil is more porous and sandy, enabling water to drain away and air to penetrate more easily, do the roots and living organisms extend deeper.

### *Making a profile diagram*

This layering, or *stratification*, of jungle vegetation is generally rather difficult to observe. To some extent it is visible from a low-flying airplane. But it is most obvious where the jungle can be seen in profile—where a clearing or a road has recently been cut into the jungle, for example. Even here, the view of the interior structure of the jungle is short-lived.

As soon as light is allowed to penetrate, the new growth of vines and trees at the jungle edge soon masks the view of the interior.

Yet if we are to obtain a clear picture of the architecture of the jungle, we need facts and figures about such things as tree heights and the openness or density of the stories. A good way to gather such information is by making a clearing in the forest and then systematically measuring the exact heights and spacing of all the trees. The results can be shown as *profile diagrams*, which are in effect maps of cross-sections of the rain forest drawn to scale.

The botanist begins preparing his profile diagram by selecting a narrow sample strip of jungle. To obtain a true picture of the interior structure of the rain forest, the sample must be selected well away from a clearing or a stream, where the increased light causes new growth that would obscure the pattern of stratification. Although the exact size of the sample strip is not very important, an area 25 feet wide and at least 150 feet long seems to give a fairly representative picture, and experience has shown that a strip at least 250 feet long is even better.

Once he has selected the area for his profile and marked its boundaries, the botanist draws a ground plan showing the positions and diameters of all the trees in the sample area. And then the difficult part of the job begins, for now it is necessary to make a set of further measurements on each tree, including its total height, the height to the lowest branch and the lowest leaves, and the width of the crown. Obviously, it would take a long time to make these measurements on every single tree in the sample strip. To simplify the task, the smallest saplings, say those less than fifteen feet high, are usually left out.

For trees up to about thirty feet tall, all the necessary measurements can be made or accurately estimated with the help of a tape measure and a pole of known length. But for taller trees the task is usually more difficult. When the crown of the tree is clearly visible, it is a simple matter to calculate its height by using a surveyor's level to find the angle to the top of the tree and measuring the distance along the ground from the observer to the base of the tree. If the top of the tree cannot be seen, enough of the surrounding trees must be cleared to make it visible.

If this cannot be done, the tree itself must be felled. Even

Although cauliflory is seldom observed outside the tropics, the redbud of eastern North America frequently produces cauliflorous blossoms on its trunk and larger branches. The closely related European redbud, or Judas tree, often planted as an ornamental, is another nontropical example of a cauliflorous tree.

Most forests are stratified, although the pattern of stratification differs in each one. In deciduous forests in the eastern United States, the tallest trees, such as maples and beeches, form an almost unbroken canopy. Their foliage shuts off much of the light from the understory of small trees such as dogwoods, and from the rather sparse shrub layer and herb layer near ground level.

here there are often complications. The tree may be so firmly bound to its neighbors by vines that it will not fall even when its trunk has been completely severed. Or the crown may break so that an accurate measure of the height is impossible.

In practice, one can usually get a fairly good idea of the heights of all the trees on a strip by sighting with a level, by felling, or by estimation from the heights of already-measured neighboring trees. In addition, the botanist keeps a record of the species of each tree. If the tree is unknown, leaves and, if possible, fruits and flowers are collected for later identification. Records of such things as the presence of epiphytes and vines also are kept.

All this is recorded on the profile diagram, adding height, shape, and so on for every tree along the sample strip. When this has been done, quite a clear picture of the jungle's interior structure emerges.

### *Layers of life*

No two profile diagrams ever look exactly alike. Even the number of layers, or strata, may be different in different areas. Depending on environmental conditions, the rain forest may include three, four, or five layers of vegetation. Nevertheless, by comparing profile diagrams from different places, we find that there is a kind of pattern in the architecture of the jungle, and the pattern often remains more or less the same for mile after mile. Even jungles as far apart as those in Amazonia and Borneo may be built on very similar plans.

The most significant feature, of course, is the fact that they *are* stratified, just as forests elsewhere are stratified. In temperate forests in the northeastern United States, for example, there is usually a canopy of the tallest trees, an understory of shorter trees, a shrub layer, and an herb layer on the ground. But in these forests the canopy is usually the densest layer, shutting out most of the light from all the lower stories.

Although the strata in the rain forest are fairly well defined, there are no sharp breaks between stories. For one thing, each of the lower stories includes some young trees still growing upward to higher levels. Each layer, moreover,

## RAIN-FOREST STRATIFICATION

Although the stratification of jungle vegetation differs by area, this profile of a South American jungle represents a widespread pattern of rain-forest stratification. The A-story is composed of widely spaced, umbrella-shaped crowns of the tallest trees. In contrast to deciduous forests in the eastern United States, this tallest layer is open and airy. In the B-story, the mop-shaped crowns of medium trees also are fairly far apart. The densest layer of vegetation is formed of the narrow, closely spaced crowns of short, C-story trees. Except for scattered sunflecks, little light passes through this layer of foliage to the sparse vegetation of the D-story and E-story nearer to ground level.

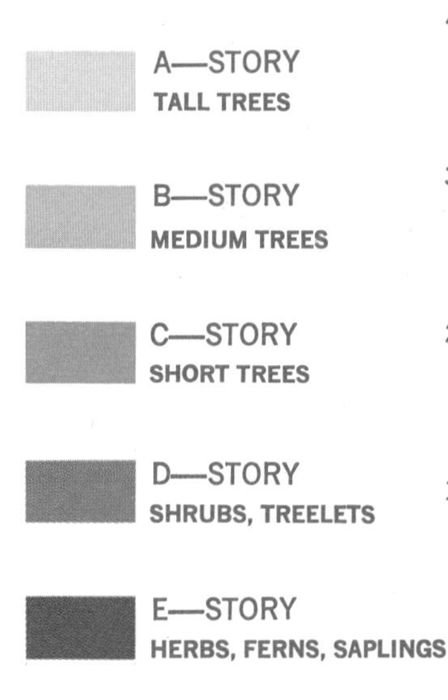

includes trees of many different species. No two of them are exactly alike in the height to which they grow at maturity, the shape of their crowns, or anything else. The architecture of the jungle, as a result, is very irregular and must not be compared too closely to a building.

Yet this stratification is of utmost importance to all the jungle's inhabitants, for it literally provides the framework of their lives. The living conditions for small plants growing near the ground, for vines and epiphytes, and for most of the jungle animals result from the stratification of the trees. Even the more mobile inhabitants of the jungle, such as birds and mammals, cannot be divided simply into those of the treetops and those of the undergrowth, since many live chiefly in the middle stories.

The vertical distribution of birds and mammals, in fact, is controlled by a variety of factors, including the types of food available at different levels and the opportunities for different types of locomotion, such as running, climbing, gliding, and flying. The stratification of insect life depends on these same factors, as well as on differences in the light, temperature, and humidity at different levels.

**Enjoying a free ride, a young pangolin, or scaly anteater, accompanies its mother as she climbs a huge tree. In their constant quest for ant and termite nests, pangolins regularly travel between ground level and the upper layers of the jungle canopy.**

### *The jungle makes its own climate*

Because of the complicated architecture of the jungle, environmental conditions inside the forest differ not only from those in a large clearing outside the jungle; they also differ at every level within the forest itself, from ground to canopy. The changes in factors such as light, temperature, and humidity, at various levels, moreover, are different on dry days and wet days and different by night and day. Each level, in fact, has its own *microclimate*, or climate in a small space, within the larger climate of the jungle as a whole.

You can most easily observe the effects of the forest on microclimate by stepping into the rain forest from a small clearing where a native farmer is growing a crop of corn or manioc. The change from heat and comparative dryness

**Gorillas, the largest of the great apes, spend most of their time on the ground. But despite their great size—old males may weigh 450 pounds or more—they sometimes climb to the lower levels of jungle trees to feed or sleep.**

in the open to the cooler, moister conditions inside the forest is amazing. The contrast is far greater than that between conditions inside and outside a temperate forest because in the jungle the trees are taller and there are more layers of leaves between the ground and the open sky. Since the jungle is an evergreen forest never bare of leaves, this effect is felt all year round.

The most obvious effect of the trees is greatly to reduce the amount of light reaching the undergrowth. Except in the sunflecks that move about and change in size throughout the day, the light a few feet above the ground is less than one percent of full daylight. The trees, moreover, do much more than merely reduce the intensity of the light. Some light is filtered through the leaves, perhaps more than once, and some is reflected from their shiny surfaces or from bark of various colors. Thus, as sunlight passes down from story to story, some of the wavelengths that make up white light are absorbed more than others. As a result, the light that eventually reaches the undergrowth has lost much of the blue and green wavelengths and includes proportionately more red and infrared light.

The radiation of invisible heatwaves to lower levels also decreases rapidly. In the middle of a sunny day, the temperature in the canopy might be 90 degrees, while in the undergrowth it is only 80 degrees. The difference is less on cloudy days, but even so, the maximum daytime temperature in the canopy is always higher than in the undergrowth.

After dark, especially on clear starlit nights, the situation is just the opposite. Since heat is radiated more readily from the treetops, the air is generally cooler in the canopy than near the ground. The difference between day and night temperatures is thus considerably greater in the upper stories than in the lower ones. The upper stories are hotter by day and cooler by night while in the undergrowth neither temperature nor humidity varies greatly between night and day or between one day and the next.

Beneath the soil, conditions are even more stable. At ground level the temperature may fluctuate over a range of two or three degrees, but at a depth of about three feet

**The dimness of ground level in a South American jungle is brightened only by the sunflecks that filter through gaps in the canopy. Yet even this meager ration of light is sufficient for many kinds of shade-tolerant plants.**

below the surface, the soil temperature is about 77 degrees and never changes at all.

Besides acting as filters and barriers to radiation of light and heat, the trees also interfere with air movements of all kinds. The climate of the wet tropics is generally less windy than other parts of the world. But in the undergrowth, even this small amount of air movement drops to almost nothing except during brief squalls and just before thunderstorms. Through most of the day small bits of paper never move and smoke rises quite vertically. In open clearings, by contrast, leaves are constantly ruffled by gentle breezes, to the despair of anyone trying to take close-up photographs.

This stillness of the air is largely responsible for the constant wetness of the undergrowth, for in the absence of breezes, evaporation proceeds very slowly. Usually the air is so nearly saturated with moisture that it is difficult to dry anything without the help of a fire.

The extraordinarily constant conditions near the ground suit some organisms and not others. Except for the rather feeble illumination during the day, conditions at ground level are somewhat similar to those in a cave. As it happens, both caves and the litter and rotten wood on the jungle floor harbor certain small animals that are regarded as living fossils, primitive creatures that seem to be relics from some distant geological period. Possibly they owe their survival to the distinctive never-changing living conditions found in both habitats. One of the most remarkable of these relics is *Peripatus*, a curious, soft-bodied centipedelike creature that resembles a worm with legs. Several species of *Peripatus* live in litter and rotten wood in both the Old and New World jungles.

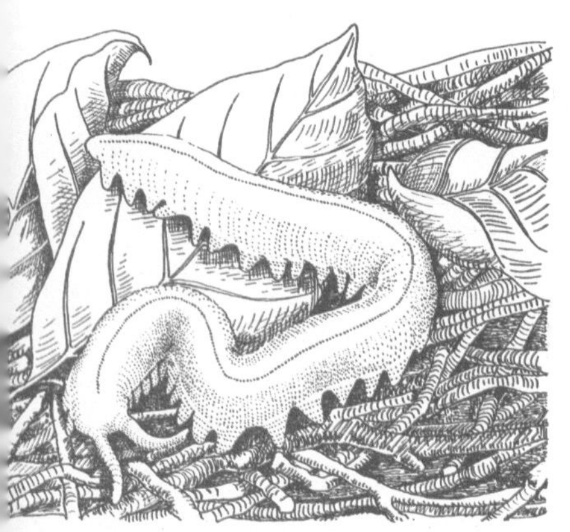

One of the strangest dwellers in the never-changing, constantly humid microhabitat under leaf litter and rotting logs on the jungle floor is *Peripatus*. Several species of these centipedelike creatures live in both Old and New World jungles. They generally remain in hiding during the day and come out at night or during rainfalls to feed on dead animal matter.

### *A place for everything*

All in all, it is clear that the jungle includes many different microclimates, with living conditions changing gradually as you proceed from one level to the next. Some forms of life are adapted for existence at more than one level. The tall trees and lianas, for example, begin their lives in the dense shade and high humidity of the undergrowth. As they grow to maturity, they gradually pass into the bright light and varying temperature and humidity of the canopy. Thus it is not surprising that young plants of such species often have

Just as the plants of the jungle are stratified, different kinds of animals also live in more or less restricted vertical ranges. Boundaries between animal zones are rather indefinite, however, since animals can easily climb, leap, glide, or fly from level to level. Yet each layer, from the highest canopy to the jungle floor, tends to have its characteristic animal inhabitants. Shown here are some of the mammals that one researcher observed at various levels in the Malayan jungle. Different species of birds, insects, amphibians, and other kinds of animals also tend to live at particular levels in the jungle.

**Tayras, bushy-tailed relatives of our familiar weasels, inhabit tropical rain forests in Central and South America. Nimble and fast, they hunt for birds and small mammals both on the ground and high up in the trees. In emergencies they will also dine on fruits and honey.**

leaves that are quite different in shape and texture from those of the mature plants.

Very few kinds of animals, on the other hand, are equally at home at all levels in the forest. Most of them remain pretty much in their own vertical range, although some, especially small insects such as mosquitoes, change their level at night or move up and down with changes in the weather.

William Beebe was one of the first naturalists to study the height zonation of animals in the tropics. He divided the jungle at Kartabo, Guyana, into several height zones similar to those described above and listed the chief types of birds and mammals in each. On the ground he noted such animals as partridges, tinamous, armadillos, deer, peccaries, tapirs, and various rodents. For the low jungle—up to twenty feet—he listed antbirds, manakins, wrens, and opossums. In the more open midjungle, between twenty and seventy feet, he found birds such as hawks, owls, honeycreepers, and pigeons, and mammals such as bats, marmosets, squirrels, and coatis.

The treetops, above seventy feet, were inhabited mostly by animals such as cotingas, macaws, parrots, and various kinds of monkeys. Beebe also noted that some of the animals ranged over more than one zone. The tyrant flycatchers, for instance, occurred in both the treetops and the midjungle,

Like tayras, ocelots range through all levels of the jungle in pursuit of prey. These skillful climbers hunt by night, stalking any small mammals, birds, or reptiles they can find. Relentlessly hunted for their fashionable fur, these sleek tropical American cats are rapidly dwindling in numbers.

while he found hummingbirds at all levels from the undergrowth to the treetops.

As might be expected, there is a close link between the kinds of food available at different levels and the feeding habits of the animals that live there. The monkeys of the treetops feed mainly on fruits, flowers, and leaves. The larger birds of the treetops, such as parrots and toucans, also eat fruits, especially nuts which they crack with their powerful beaks. Flying insects are much more plentiful in the treetops and middle jungle than among the closely packed foliage and twigs lower down. Hence it is mostly in these levels that many insect-eating birds and mammals feed. Many of the vegetarian ground dwellers, such as rodents and peccaries, in turn depend on the shower of fruits, flowers, and leaves that continually fall from the treetops.

Another scientist who studied the animals of the Malayan jungle found similar relationships between height zones and feeding habits. As in Guyana, the highest canopy is inhabited mainly by leaf and fruit eaters, such as gibbons, leaf monkeys, giant squirrels, flying squirrels, tree rats, slow lorises, and flying lemurs. The middle level is the realm of birds and bats that catch flying insects, and mammals such as monkeys, civets, tree squirrels, tree rats, and tree shrews, which have a mixed diet and can climb about freely. Some of the middle-level animals, such as the clouded leopard, spend part of their time in the trees and part on or near the ground. Finally there are the animals that keep mainly on the ground and seldom climb or fly very far.

Some of the smaller birds and mammals at ground level are *carnivores*, or flesh eaters; some are *herbivores*, or plant eaters; and some are *omnivores*, or mixed feeders. The larger mammals of the forest floor, including such huge creatures as elephants and rhinoceroses, as well as deer, pigs, porcupines, and the Malayan bear, on the other hand, are exclusively herbivorous.

### *Insect stratification*

Insects and other invertebrates, as well as other small creatures such as lizards and frogs, also are zoned according to height. But less is known about their distribution, since they are not as easily observed as the larger animals. Their distribution can be studied only by careful trapping at different levels.

One group of insects which has been carefully studied is mosquitoes. In the African jungle they include high-level types which bite monkeys, birds, and other treetop animals, and low-level types that live in the moist, shady undergrowth. The high-level group is very important, for it includes species of *Aedes*, the mosquitoes that transmit the virus of yellow fever. And, as in tropical America, monkeys act as a reservoir for the virus; there is always a risk that the infection will spread from the monkeys to human populations.

Equally dangerous to man are the low-level species, for they include *Anopheles* mosquitoes, the type that transmits malaria. In tropical America, *Anopheles* mosquitoes are found at high levels as well as low, and since the water "tanks" of epiphytic bromeliads provide one of their chief breeding places, malaria control in these jungles is especially difficult.

Like the mosquitoes, the butterflies of the treetops are different from those of the undergrowth. The splendid blue morphos of tropical America stay mostly in the treetops. They are seldom seen near ground level except in sunny clearings and along riverbanks. In the shady undergrowth you find instead the delicate, slow-flying *Heliconius* and "leaf

Any visitor to the tropical rain forests of South America is certain to notice the many kinds of *Heliconius* butterflies that live here. Characteristically patterned with vivid combinations of red, orange, yellow, white, and black, these handsome insects usually fly close to ground level, threading their way through the twigs and foliage.

butterflies." The latter, when they rest on a branch, show only the drab undersides of their wings, which look so much like dead leaves that they become almost invisible.

## *Many ways of life*

The plants and animals of the tropical jungle not only live at particular levels but also live in particular ways. In a forest or a prairie, or in any other natural community for that matter, each organism plays its characteristic role in the working of the community as a whole, in much the same way that members of a human society contribute to the life of their city or nation. In the language of *ecology*, the science studying the relationships of living things to each other and to their environment, the role of an organism in a natural community is called its *niche*.

As Charles Elton, one of the English pioneers of ecology, put it, every animal in a natural community has an "address" and a "profession." Its address is the place where it carries on its activities, such as in the forest canopy or under the bark on fallen logs. But its profession, or niche, might be carnivore, fruit eater, destroyer of dead wood, and so on.

Animals are often remarkably specialized in their professions, such as herbivores that eat only the leaves of one kind of tree. The three-toed sloth of tropical America, for instance, feeds almost exclusively on the leaves of the cecropia tree.

The term carnivore, in turn, includes animals as different in size, food preferences, and habits as tigers and tiny insect-eating shrews. It also includes birds that catch insects on the wing, as swifts and nightjars do, as well as woodpeckers which seek insects and their larvae under bark. Insects such as mantids which prey on smaller insects also are carnivores. Thus many different niches are covered by such general terms as carnivore, herbivore, and mixed feeder.

It should be noted, too, that we were actually discussing niches when we mentioned certain animals of the Old World jungles that replace others found only in the New World.

**Broad petal-like limbs and body provide perfect camouflage for a flo ver mantid in the Borneo jungle. These amazingly specialized insect predators are easily mistaken for blossoms, both by the smaller insects on which they feed and by creatures that prey on mantids.**

## THE MOUSE OPOSSUM: AERIAL OPPORTUNIST

The mouse opossum, small enough to fit in the palm of a man's hand, is one of several species of opossums found in South American jungles. This nimble dweller in the treetops, equipped with a long, naked, prehensile tail, is an omnivore, ready to take advantage of any food source it encounters. Insects are a mainstay in its diet, and it seems undaunted even by a giant grasshopper

nearly as big as itself. But the mouse opossum just as willingly nibbles on the fruit that abounds in the jungle. Around banana plantations it is even considered to be a nuisance. And on occasion it adopts the role of scavenger, dipping its paws into the tanks of epiphytic bromeliads to fish out the bodies of insects and other creatures that have drowned in these aerial reservoirs.

The roots of this strangling fig have all but enveloped the trunk of a mahogany in a grip of death. In time, the strangler, which began life as an epiphyte on the mahogany's branches, will probably kill its host. Eventually the mahogany will rot away and the strangler will stand as an independent tree, supported by a hollow cylindrical trunk formed of the network of roots that engulfed the host tree.

nearly as big as itself. But the mouse opossum just as willingly nibbles on the fruit that abounds in the jungle. Around banana plantations it is even considered to be a nuisance. And on occasion it adopts the role of scavenger, dipping its paws into the tanks of epiphytic bromeliads to fish out the bodies of insects and other creatures that have drowned in these aerial reservoirs.

In fact, they occupy corresponding niches. The address of toucans is tropical forests of America and that of hornbills is tropical forests of Asia and Africa. But the profession of both is bird that eats mostly fleshy fruits, plus nuts, and occasionally small animals. The flower-visiting sunbirds of the Old World and hummingbirds of the New World also occupy similar niches.

The same thing is true of plants. Tall trees, small shade-tolerant trees, vines, epiphytes, and so on each occupy particular niches in the jungle community. And, as with animals, there are many examples of plants in Old and New World jungles which, though they are different species, live in much the same way. The parasol tree, for example, is a fast-growing, short-lived tree that characteristically springs up in gaps and clearings in the African jungle. In tropical America, the cecropias occupy a closely corresponding niche.

Among the most spectacular of all lianas are the climbing palms or rattans of Asian jungles. Their stems are often thorny and can grow to a length of several hundred feet, looping from tree to tree across the forest canopy. The detail shows the characteristic leaves and blossoms of one of these unusual vines, which are widely used for weaving baskets and wicker furniture.

## *Plants that climb*

It would be useful now to take a closer look at some of the more unusual plants that contribute to the distinctive architecture of the jungle. Certainly one of the most striking features of rain forests everywhere are the woody vines, or lianas, that rise up to the canopy like great ropes and hang down in enormous loops, linking one story to the next. Their West Indian name, bush rope, is particularly appropriate, for their stems are so strong, yet flexible, that a man can climb them quite safely. Brazilian tree climbers often swarm up lianas as easily as sailors climb the riggings of a ship.

When they are young, lianas look very much like the seedlings or saplings of trees. And like many of the tall trees, their leaves at this stage when they are still growing in the dense shade of the undergrowth often are very different from the leaves produced when the plant reaches full sunlight in the treetops. The lianas do not begin really vigorous upward growth until the death of a branch or a whole tree creates a gap in the canopy. And then they begin to shoot up rapidly, often attaching themselves to young trees and growing upward as the tree grows. Different species climb in different ways. A few cling to trees by means of aerial roots; some twine like morning glories and others have hooks or coiling tendrils like those on grapevines.

A snarl of lianas traces a pattern of loops and knots against the sky in a Brazilian jungle. These enormous woody vines, some of them hundreds of feet long, are a characteristic feature of all tropical rain forests.

Normally the stems branch very little until the leaves are fully exposed to the light. But once they reach the canopy level, they often branch so vigorously that they compete seriously with the trees for available light. They may even form masses of foliage that are heavy enough to bend and break small trees. As they spread across the treetops, their stems can become very long. Lianas over two hundred feet long are not at all uncommon. And in Java, one of the climbing palms or rattans which are so characteristic of Asiatic jungles is reported to have reached a length of eight hundred feet.

So long as the jungle remains undisturbed by felling or windthrows, the lianas and trees usually compete on more or less equal terms. But as soon as the trees are damaged in any way, or large gaps are formed, the increased light causes an outburst of growth and a dense tangle results. Trees whose crowns are broken by tornadoes, for instance,

The roots of this strangling fig have all but enveloped the trunk of a mahogany in a grip of death. In time, the strangler, which began life as an epiphyte on the mahogany's branches, will probably kill its host. Eventually the mahogany will rot away and the strangler will stand as an independent tree, supported by a hollow cylindrical trunk formed of the network of roots that engulfed the host tree.

become covered with dense blankets of lianas that may eventually kill them. And on abandoned clearings, vines of all kinds grow so luxuriantly that the secondary forests that eventually spring up in the clearings can usually be recognized as such by the profusion of lianas.

### *Plants that strangle*

Another interesting group of jungle plants are the stranglers, which are quite unlike any plants found in temperate forests. A strangler begins life as an epiphyte on the branch of a large tree, where the seed has been dropped by a bird or a fruit bat. As it grows, the young plant begins to form roots of two different sorts. Some dangle freely in the air, while others creep down the trunk of the host tree.

A giant among stranglers is the famous banyan, a strangler fig of southeast Asia, which rapidly grows to enormous size. In addition to strangling roots that envelop the trunk of the host tree, hundreds of aerial roots grow down from the branches. Some of these thicken and form stout supports for the tree. Banyans frequently grow so large that, in Indian villages, a single tree may shelter an entire open-air marketplace.

By the time an aerial root reaches the ground, it may be ninety feet long or even longer. When it finally touches the soil, the root branches and forms a solid anchorage. The rest of the root shortens slightly so that it becomes taut like a stretched wire, and gradually thickens and becomes more and more woody. Now that the plant can obtain water and nutrients from the soil, it quickly develops a sizable crown and begins to compete seriously with the host tree for light and space.

Meanwhile the roots surrounding the host's trunk have continued to grow until now they are an inch or more thick. Gradually these roots become joined here and there, so that a tough woody network begins to encase the host's trunk. The network may eventually become so dense that it strangles the tree and kills it. When this happens, the trunk of the host rots away, leaving the strangler as an independent tree with a trunk composed of a hollow cyclindrical network of woody roots. Stranglers do not invariably kill their hosts, of course, but they always harm them by taking some light from the crown.

The commonest stranglers are species of figs (related to the edible fig) and they are numerous in the jungles of both the Old and New Worlds. The clusias, tropical American plants with handsome pink or white flowers, have adopted a very similar mode of life, as have a few plants belonging to other families.

## *Plants that grow on other plants*

In the complicated architecture of the jungle, there is practically no living space that is not used by life of some sort, as the epiphytes, or air plants, make clear. As we shall see, there are even animals that live only in these plants that live on other plants.

Epiphytes grow at all levels in the jungle. A few, such as the tiny filmy ferns, are found in the deep shade of the undergrowth, but the majority grow high up in almost full daylight. And they seem to grow everywhere: attached to the trunks and branches of trees, on the stems of lianas, and even on the leaves of larger plants. Moreover, they grow in such variety that it is not unusual to find thirty or more species on a single tree.

As in many temperate forests, mosses, liverworts, and lichens grow as epiphytes. But in the jungle they are not very conspicuous and are far surpassed in quantity and variety by ferns and flowering plants. Throughout the tropics, the most important families of epiphytes are the orchids and the arums, while in the American tropics there are many epiphytic cactuses and bromeliads as well. The bromeliad, or pineapple family, in fact, includes one very familiar epiphyte that extends outside the tropics. The well-known Spanish moss that festoons trees in the southeastern United States is in reality a flowering plant related to the pineapple.

Wherever they live, epiphytes are always faced with the problem of securing sufficient water for growth. Since they grow on trees rather than with roots in the ground, they must have some way of finding a water supply to last them from one rainfall to the next. Even in the "ever-wet" rain forest, this may involve a wait of as much as three weeks. They also have a nutrient problem. Unless they can survive on the meager ration of nitrates and other minerals contained in rain water, they must find additional sources of nutrients. Different epiphytes have solved these problems in different ways.

If you pull a large bromeliad from a branch, you are quite

**In the jungle, epiphytes grow everywhere. Some, such as this colorful bromeliad (*left*), perch on branches. Others, such as elkhorn ferns (*opposite page, left*), cling to tree trunks. Even the twisted stems of lianas provide support for epiphytes (*opposite page, right*).**

*Bletia catenulata* (left), *a native of South America, is one of fifty species of Bletia orchids. Most are terrestrial and, like this one, bear their blossoms on long wandlike stems.*

## JEWELS OF THE JUNGLE WORLD

**Among the most colorful, varied, and sought-after of all flowering plants are the orchids. This huge family, including some 24,000 species, ranges throughout the world, with orchids blooming everywhere from arctic bogs to the depths of tropical rain forests. Whereas most orchids of temperate climates spring from terrestrial roots, in the tropics where orchids are most abundant, epiphytic species predominate. Wherever they grow, orchids come in an astonishing variety of shapes, colors, and sizes, many with tiny white or greenish blossoms no more than an eighth of an inch across, others with splendidly colored petals several inches long.**

*The scorpion orchid, Arachnis flos-aeris* (right), *is a vinelike plant that forms tangles over bushes and trees in low-lying jungles in Malaya, New Guinea, and elsewhere in the East Indies. Its somewhat sinister-looking four-inch flowers have made it a favored plant for cultivation.*

*Seventy species of Phalaenopsis orchids, most of them epiphytic, grow in southeast Asian jungles. Many of them, like the white moth orchid* (left), *are cultivated by orchid fanciers.*

likely to receive a gallon or so of dirty water on your head, for these plants collect both water and nutrients in tanklike reservoirs. Bromeliads have short stems surmounted by a rosette of narrow leaves. In epiphytic species, the leaves usually overlap at their bases to form watertight cups in which rain water collects. In addition, many fragments of dead leaves and flowers as well as dead animal matter fall into the tanks or are carried there by ants and are broken down by microorganisms into mineral nutrients.

Curious little scales on the inner surfaces of the leaves enable the plants to absorb the water and dissolved nutrients they require. Some bromeliads can obtain water in no other way: the roots have lost the power of absorption and serve only to attach the plant to the tree. This system for collecting water is so successful that some species of bromeliads flourish even on telephone wires where their sticky seeds are often deposited by birds.

Many other epiphytes rely mainly on water stored in living tissues of the plant. Most epiphytic orchids have swellings on the stems or at the bases of the leaves in which considerable amounts of water can be stored. In dry weather when the store of water runs low, these *pseudobulbs*, as they are called, become thin and shriveled. In the West Indian mistletoe, *Rhipsalis*, which is actually a species of cactus, the whole plant is fleshy and contains a store of water.

Certain other epiphytes have special water- and humus-collecting structures almost equal in efficiency to those of the bromeliads. The elkshorn fern, *Platycerium*, is called a bracket epiphyte because, in addition to its normal fernlike leaves, it has special bracket leaves that are pressed against the bark of the tree in the form of brackets. Rain water and humus collect in the pockets between the bracket leaves and trunk, and the roots of the fern grow into the pockets. Even more remarkable is *Dischidia rafflesiana*, an epiphytic flowering plant of the Malaysian jungle. *Dischidia* also has two kinds of leaves, one of which grows in the form of an oval sac about three inches long. This also forms a pocket for collecting dead animal and vegetable material which is absorbed by roots growing into the sac from the stem.

**The epiphytic fern, *Drynaria*, makes its own soil. In addition to fernlike foliage leaves, it has short bracket leaves, most of them brown on this specimen. Debris collects behind these leaves and decomposes to form humus.**

## *Of ants and epiphytes*

Perhaps the strangest looking of all epiphytes are two plants called *Myrmecodia* and *Hydnophytum* which are quite common on trees in Malaya and Borneo. In both, the body of the plant grows in the form of a large tuber rather like a turnip in size and appearance. At its lower end, each of these tubers is anchored to the tree by thin wiry roots, and at the top there is a short leafy stem.

If you cut one of the swollen tubers in half, you will find a mass of whitish tissue honeycombed with tunnels and enclosed in a corky outer layer. The tunnels almost invariably are inhabited by small fierce ants. Whether the ants benefit or harm the plants has never been proven. They may be merely guests that find the tunnels a convenient place in which to live.

There is in fact a close association between most epiphytes and ants. The ants sometimes live in special parts of the plants, such as in the tubers of *Myrmecodia* and *Hydnophytum*, or in the sacs of *Dischidia*. Or, as orchid collectors know only too well, they may live among the roots of many other species. In some cases the advantage of this relationship to the plant is not clear, though the ants certainly assist many epiphytes by bringing back leaves, petals, seeds, and other material from their foraging expeditions. This plant material rots down to form humus and provides the plants with a sort of soil that holds a certain amount of water as well as providing nutrients.

It would seem, then, that the rain forest is a remarkably well-ordered world. Although at first glance the jungle seems to be a totally disorganized chaos of superabundant greenery, closer examination shows that it has a rather definite structure. Tall trees, short trees, vines, and epiphytes each have a specific role in the general scheme of things. And animal life in turn is well fitted into the overall architecture of the rain forest. In the jungle, in short, there is a place for everything—and everything remains pretty much in its place.

**In the jungle, every conceivable bit of living space seems to be occupied. Even the slender stem of a dangling liana has provided sufficient foothold for an epiphytic aroid to sprout and produce a graceful bouquet of greenery.**

# A World in Harmony

Jungles are among the oldest natural communities on earth. Fossil plants discovered in Indonesia and other parts of the tropics show that jungles have existed in more or less their present form for perhaps sixty million years.

During this incredibly long period of time the world's climate has gone through many changes. In moist periods, tropical forests spread into areas that are now deserts, and in dry periods they may have occupied smaller areas than they do now. Yet, through it all, the plants and animals of the jungle have continued to survive and flourish, day in and day out, year after year.

How has the tropical rain forest managed to maintain itself over such a long period of time? For one thing, each member of this vast and varied assemblage of plants and animals fills its own particular niche in the community as a whole. At the same time, every organism affects other organisms in many ways. Trees and other green plants, for example, compete with one another for light and living space, and the plants themselves are eaten by hosts of herbivorous insects, mammals, birds, and other animals. But the plant eaters in turn are eaten by other animals. And when they die, both plants and animals nourish a huge population of fungi, bacteria, and other small organisms.

The interrelations are not always struggles between antagonists, however. The trees not only compete with each other for space and light; they also provide the shade which is essential for the survival of many smaller plants. And they provide many kinds of animals with nesting sites and concealment from their enemies. Then, too, there are more complex relationships, such as those between flowers and their pollinators and between plants and the animals that help to disperse their seeds.

But what about the jungle as a whole? This vast assemblage of plants, animals, and their nonliving environment makes up what is known as an ecological system, or more simply, an *ecosystem*. How does the jungle ecosystem work? How does it keep in balance?

### *Energy from the sun*

In the jungle, as everywhere, the motive power of the system is the sun's energy. The only organisms able to make direct use of sunlight, however, are green plants. In a series of complex chemical reactions, they use the energy in sunlight to rearrange the molecules of water and carbon dioxide into sugars and other carbon compounds. This process of *photosynthesis* takes place only in the presence of *chlorophyll*, the green coloring matter in plants.

In addition, the plants need compounds of nitrogen, phosphorus, potassium, iron, and other elements for their growth. They obtain these mineral elements in the water they absorb from the soil. Together with the carbon compounds built up in photosynthesis, these substances are used to make amino acids. And amino acids, in turn, are the essential building blocks in the production of proteins, the basic constituents of all living things.

For this process of converting solar energy into usable food, tropical jungles, of course, have a basic advantage over forests in other areas. They are located at or near the equator, the part of the earth that receives the greatest amount of solar radiation. In the temperate zones, well to the north

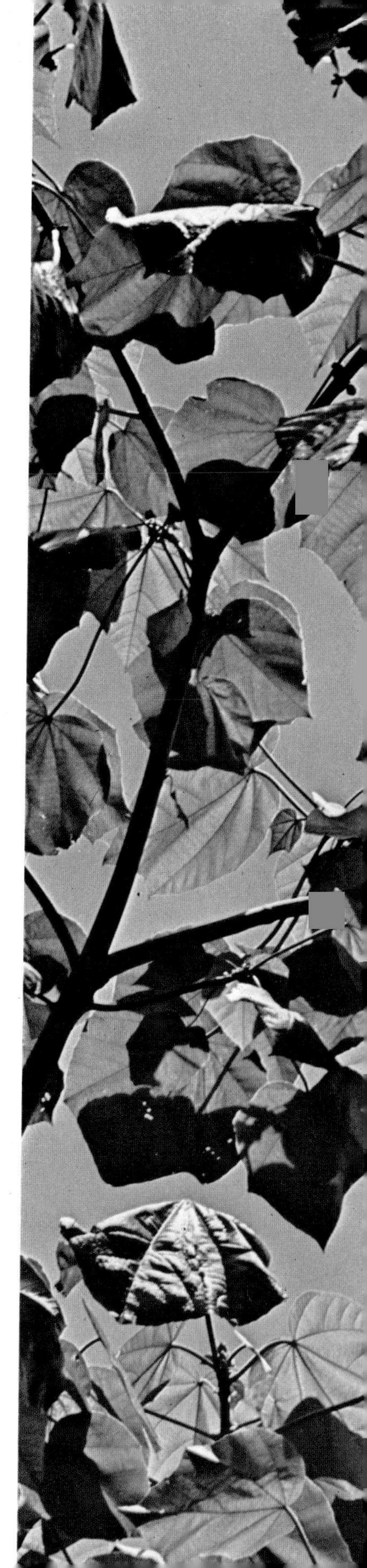

**The green leaves of a balsa tree form a colorful mosaic against the sky in a jungle clearing in Puerto Rico. Within the chlorophyll-laden cells of all green plants, the energy of sunlight converts water and carbon dioxide into the basic foods upon which all life depends.**

and south of the equator, the sun's rays strike the earth from different angles at different times of year. In temperate zones, the summer sun shines down from almost directly overhead, but in winter it shines on the earth from a position much closer to the horizon. When the sun's rays shine from this lower angle, the energy in each sun ray is spread over a larger area of the earth's surface and is filtered through a much thicker layer of atmosphere.

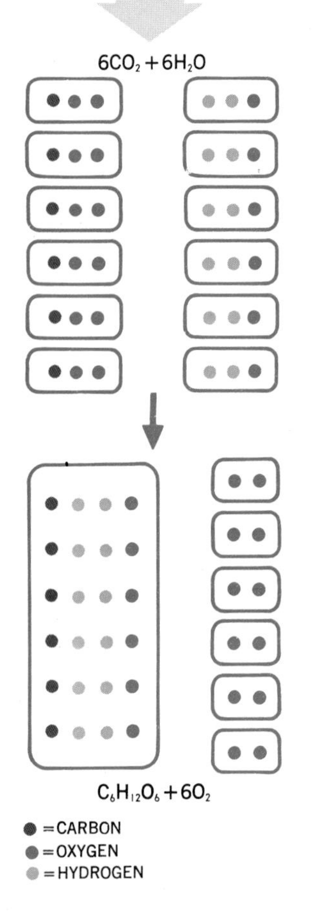


In photosynthesis, plants use energy from the sun, carbon dioxide ($CO_2$), and water ($H_2O$) to make simple sugars ($C_6H_{12}O_6$). In the process, excess oxygen ($O_2$) is released into the atmosphere. The diagram shows how atoms of carbon, oxygen, and hydrogen are rearranged in this complex process.

In the tropics, on the other hand, the sun's rays beat down from a high angle—more or less directly overhead—throughout the year. As a result, tropical jungles have much more energy at their disposal than temperate forests do. To be exact, a square centimeter of horizontal surface in the tropics receives over one hundred kilocalories of solar energy per year. (A kilocalorie equals one thousand calories; calories are the basic units for the measurement of heat energy.) This is more than double the annual amount of energy that falls on a square centimeter of surface 60 degrees north of the equator.

Plants of the rain forest, moreover, are able to go on producing food at a steady rate throughout the year. In temperate zones, the smaller amounts of energy available in the cold winter months are largely unused, since the activities of the plants slow down or cease entirely in winter.

For several reasons, plants vary in their ability to use the available energy. Factors such as the amount of water and essential nutrients in the soil affect their food-producing ability. And even under ideal conditions, plants are never very efficient. At best they can use only one or two percent of the available energy for food production. Despite these limitations, the amounts of organic matter they produce is astonishing.

## *The food factories*

An acre of tropical forest may include anywhere from 150 to 300 trees a foot or more in diameter. In addition, there are many smaller trees and saplings and an almost uncountable number of vines, epiphytes, and small herbaceous plants. And every one of them is, in a sense, a factory for the production of food.

To comprehend the great amounts of food they actually

produce, it would be interesting to know first of all what *biomass*, or total weight of organic material, all these living plants represent. Although we have no exact figures for any typical lowland jungle, the best estimate is about 441,000 pounds per acre. The greater part of this organic matter consists of trees, with probably about seventy-five percent in the form of their trunks and branches, a little less than twenty-five percent in the form of roots, and four to five percent consisting of leaves.

Knowing the biomass of the living plants on an acre of jungle at a given moment in time tells only part of the story, however. In a sense, the biomass of the trees and other plants can be thought of as the jungle's capital. But a more interesting question is how much new material, or "income," is being produced each year by this capital?

The question is not an easy one to answer, even if you consider only the trees and not the entire living forest. The ecologist must begin by accurately measuring the growth of trunks and branches. Then he must estimate the weight of leaves and new roots produced over the course of the year. He must also take into account losses due to such things as falling leaves and branches.

But the greatest complication arises from the fact that plants, like animals, use great quantities of food energy simply to stay alive. A simple measure of the increase in weight, or stored energy, in the trees over the course of a year would tell us only their net productivity. A measure of their total, or gross, productivity would also have to include the material consumed by the plants themselves in *cellular respiration*, the process by which energy is released through the breakdown of carbon compounds within living cells.

The rate of respiration can be calculated in various ways, but scientists usually measure it in terms of carbon dioxide loss. Plants not only use carbon dioxide when they build carbon compounds in photosynthesis; like animals, they also release it as a waste product when these compounds are broken down in respiration. Details of the techniques for measuring this carbon dioxide loss are too complicated to explore here, but the results of all these complex calculations have produced some rather startling figures on the productivity of forests in different areas.

The best estimate yet made of organic production in a tropical jungle is from measurements made in the Ivory

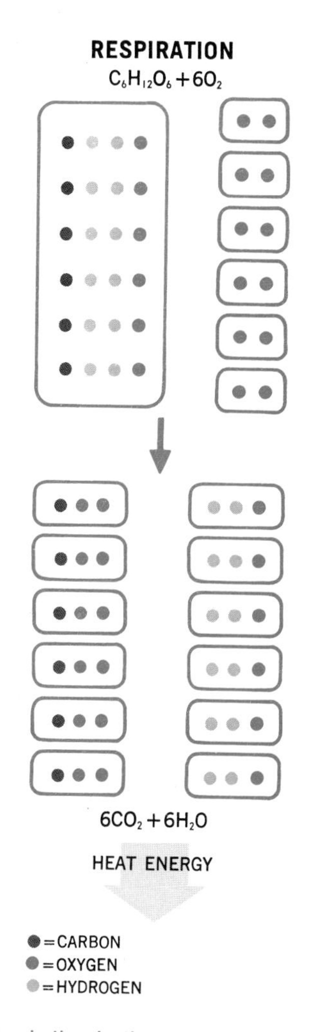


Respiration is the process by which energy stored in food substances is released to maintain an organism's life functions. In this complex series of reactions, oxygen ($O_2$) combines with simple sugars ($C_6H_{12}O_6$) and breaks them down into carbon dioxide ($CO_2$) and water ($H_2O$), releasing energy, some of which is lost as heat.

**Just as green plants are the basic producers of food, plant-eating animals are the basic consumers, using the energy stored in plant food to support their own life processes. This young tapir, munching contentedly on lush green vegetation, will lose its pattern of white spots and stripes as it matures.**

Coast of West Africa. There it was calculated that the gross yearly production of plant material—the net amount available for plant growth *plus* the amount consumed in respiration—is about 45,800 pounds per acre. This is far greater than the gross production of any temperate forest. Yet the net production—the amount actually available for plant growth—is only about 6600 pounds per acre, not very different from the net production of a beech woods in Denmark!

The tremendous difference between gross and net productivity in the jungle is due simply to the fact that the trees' rate of respiration is much greater at the high temperatures of the tropics than in cooler areas. But the similarity between the net productivity figures for tropical and temperate forests indicates that the luxuriance of jungle vegetation is not due to an exceptionally rapid rate of growth. It seems to result from the fact that the plants remain continually active, unchecked by drought or winter cold.

### *Indispensible decomposers*

Because green plants alone can transform solar energy into living material, they are called the *producers* in the jungle ecosystem. But they are only a part of the rain forest's enormous cast of characters. They share the stage with hosts of other living things, both large and small. Since none of these other living things can produce food themselves, they are all direct or indirect *consumers* of the solar energy stored up by living plants.

Some of the most important consumers are the *decomposers,* the whole army of organisms that break down the tissues of dead plants and animals. Most of them—bacteria and fungi, for example—are so small that they cannot be seen without magnification, yet they are as vital as the plants themselves to the functioning of the ecosystem.

Evidence of their work is visible on every side. Consider, for instance, the steady rain of twigs, leaves, petals, fruits, dead insects, and other material that falls continually from the canopy. By collecting this material in trays as it falls, scientists have found that the rain of litter in the jungles of Malaya and Brazil amounts to between 4400 and 7000

**Consumers in the jungle ecosystem come in many guises. These Central American dung beetles feed on droppings found beneath trees frequented by monkeys. They also lay their eggs in pellets of dung which provide food for their developing larvae.**

## COLORFUL DECOMPOSERS

The jungle, like any forest, abounds with fungi. Most are microscopic and hard to see but some, like the African species (*left*) and the Asian species (*right*), are colorful and conspicuous.

Unlike green plants, fungi are unable to manufacture their own food. Instead they obtain it by living as parasites on plants and animals or, in the case of decomposers, by breaking down the tissues of dead plants and animals. In the process of feeding on organic compounds stored in dead material, the decomposers perform an indispensable role in the functioning of the ecosystem, for they release the mineral nutrients stored in dead matter and return them to the soil where they can be used once again by green plants.

The familiar umbrellalike caps of mushrooms and toadstools are strictly reproductive organs, producing dustlike microscopic spores from which new plants will grow. The actual process of digestion and decay is performed by enzymes in the extensive networks of minute threadlike structures, called mycelia, that penetrate the dead material on which the fungi live.

pounds per acre each year—several times as much as falls in an oak or beech forest in Europe or America. Yet in the jungle less than one inch of litter usually accumulates on the ground. At tropical temperatures, decomposition takes place much more rapidly than in cooler forests. Dead limbs and whole trees that come crashing to the ground meet the same fate. Within a very short time the decomposers reduce them to masses of soft brown pulp overgrown with ferns and mosses.

One obvious benefit of this rapid destruction by the decomposers is that it keeps the jungle from being smothered in its own waste. But it is important for another reason as well. In addition to carbon compounds which nourish the decomposers, the litter and dead wood contain the mineral elements such as phosphorus, potassium, calcium, and nitrogen which are essential for plant growth. As the dead material is decomposed, mineral nutrients are released into the soil where they can be used again by plants. In this way, the inorganic elements originally derived from rock continue to circulate from the soil, to plants, to consumers, and back again to the soil.

In some cases, the transfer of the minerals back to the plants is even more direct. Many kinds of fungi live in close association with the smaller roots of trees and other plants. They cover the finer rootlets with a whitish felt and even penetrate between their cells. These associations of roots and fungi, called *mycorrhiza*, may be extremely important for the trees' nutrition, for they are apparently able to transfer mineral nutrients from decaying leaves and wood directly to the living roots.

If this is so, only a very small fraction of the soluble mineral matter would ever be set free into the soil where it could be washed away by rain water. And as it happens, the mineral content of many tropical rivers is extremely low. Some of them are almost pure rain water.

However, even the very small amount of minerals lost in rivers is usually balanced by the continuing slow breakdown

**The crystal-clear water in a jungle stream in Borneo is evidence of an ecosystem functioning smoothly. Like many jungle streams, it contains nearly pure rainwater. The vital mineral nutrients in the jungle soil are being absorbed so efficiently by plant roots that practically none are washed into the stream.**

Even predators as spectacular as South America's emerald tree boa depend ultimately on green plants for food. This one has just captured a blue tanager, probably by lying in wait among the leaves on a branch and then striking suddenly at the unwary bird. But the tanager grew fat by feeding on fruits, seeds, and insects that had fed in turn on plant material.

of rock particles in the soil. Thus the mineral circulation in the jungle ecosystem forms an almost closed cycle that can maintain itself indefinitely.

## *The flow of energy*

Like the decomposers, all the larger, more conspicuous consumers in the jungle depend ultimately on green plants for their food. Consider the jaguar, for instance. It rarely eats any plant material itself, for it is a predator that feeds on animals such as monkeys and tapirs. But these animals in turn eat leaves, fruits, flowers, and other food produced by green plants.

This interrelated series of plants and the animals that depend on them for food forms a *food chain*. In the jungle most food chains are more varied and complex than this simple plant-herbivore-predator sequence. They often include four or even five links. Occasionally the jaguar might dine on an anteater, which had fed on ants, which had fed on plants, thus creating a four-link food chain. And if the ants happened to have been a predatory species that ate other animals, the chain would consist of five links.

For the most part, the feeding habits of jungle animals are so little known that in most food chains only a few links are well understood. Even so, they seem to fit certain general rules that apply to food chains in all ecosystems. For one thing, we know that in the jungle, as everywhere else, a great deal of energy is lost at each link as food passes along a food chain. Each organism uses part of the energy in the food for its own life processes. In addition, at each link a great deal of energy is lost in the form of heat that escapes into the atmosphere.

A series of plants and animals linked by food relationships forms a food chain. In this example, the green leaves of cecropia trees store the energy in sunlight by means of photosynthesis. The sloth, unable to manufacture its own food, lives by eating cecropia leaves. The harpy eagle, a predator, feeds in turn on sloths and other herbivores, thus depending indirectly on the food originally manufactured by green plants.

Another general rule is that predators are usually larger than their prey. A jaguar is larger than a monkey or anteater, and an anteater is much larger than the ants. Yet there are exceptions. Big cats like jaguars, leopards, and tigers are so strong, fast, and skillful in stalking that they can attack and eat prey larger than themselves. Even elephants sometimes fall victim to tigers.

Carnivorous ants, because of their astonishing social organizations, are another exception. Most people have seen ants struggling with crickets or cockroaches much larger

## TOP CAT IN THE JUNGLE FOOD CHAIN

Ranging all the way from the extreme southwestern United States to northern Argentina, and thriving in drier areas as well as in jungles, jaguars are among the most widespread—and magnificent—of all New World predators. These great spotted cats, known throughout Latin America as "tigres" (tigers), grow to as much as seven and one-half feet long, and mature males sometimes weigh 250 pounds.

Large animals such as deer and tapirs are the mainstay of their diet. Around farms and ranches, they may even attack horses and cattle. But jaguars do not disdain smaller prey; monkeys, birds, fish, and reptiles all fall victim to their swift, skillful attacks. These solitary nocturnal hunters pursue their quarry just as readily through the treetops as on the ground. And they show no fear of entering the water where, it is said, they will not hesitate to attack alligators.

than themselves. And in Africa, driver ants are said to be capable of attacking and killing huge pythons if they can find the snakes when they are drowsy from a large meal.

It is also true of food chains in general that animals at the end of the chain are less numerous than those at the lower levels. This is reasonable, for the larger an animal is, the more food it needs. Obviously the jungle can support fewer anteaters than ants.

Leaf-cutting ants use their scissorlike jaws to snip neat segments from leaves and petals. Although they usually spread their leaf-cutting activities over large areas, in gardens the ants are often pests, stripping all the foliage from a single plant.

Furthermore, in most food chains, animals forming the later links tend to have less specialized feeding habits than their prey. This is so because their food is harder to obtain and their hunting cannot always be successful. A jaguar or large predatory bird, for instance, may prefer to eat deer or monkeys. But when it is hungry, it may have to content itself with rats, mice, snakes, or even insects.

Some predators, such as anteaters and pangolins, of course, are fairly specialized in their feeding habits, but truly restricted diets are much more common among herbivores. The three-toed sloth, as we have already mentioned, feeds almost exclusively on leaves of cecropia trees. And many caterpillars, bugs, aphids, and other plant-eating insects feed only on plants of one or two families, or sometimes even on just a single species.

## *Ants that grow their own food*

One of the most extraordinary examples of specialized feeding habits is that of the leaf-cutting ants, probably the most numerous and important plant-using insects in New World jungles. These ants live in colonies of several million individuals in nests that are mounds of soil several feet high and up to twenty feet in diameter.

Radiating from each nest are well-marked trails which the ants keep carefully grazed. The trails lead to all parts of the territory around the nest where the ants go to gather bits of leaves that they carry back to their underground chambers. The size of the territory varies with the size of

Seemingly oblivious of its cargo of stowaways, a leaf-cutting ant carries a piece of leaf back to the nest. There, in underground chambers, smaller worker ants, called minims, tend the gardens of fungi which the ants grow on the leaf fragments.

the ant colony; if the nest is a large one, the ants may travel as far as a hundred yards for leaves. Nor do they necessarily attack the plants closest to their nest. Most of the leaves, in fact, come from tree crowns forty feet or more above the ground. Even more astonishing is the fact, as recent observations suggest, that the foraging is carefully spread over the entire territory, with the result that no one area suffers from "overgrazing."

Long processions of leaf-cutting ants going about their business are an everyday sight in South American forests, especially in early morning. Up in the trees, the worker ants use their scissorlike jaws to snip pieces—sometimes as big as dimes—from leaves and flower petals. And then the columns of ants proceed back to the nest, with each insect

**Like an army carrying banners, a procession of leaf-cutting ants parades across the jungle floor. The ants travel along well-marked trails that are carefully cleared, guarded, and kept in repair by certain members of the ant colony. A trail-marking chemical scents the route and helps keep the burdened leaf-cutters from losing their way.**

carrying a piece of leaf or petal much larger than itself—"like Sunday-school children carrying banners," as one observer put it.

The ants are not particular as to the kinds of plants they attack. Apparently almost any leaf will do, as long as it is fresh, for the ants do not eat the leaves themselves. Their diet instead is restricted to small knoblike growths on one special kind of fungus. The ants use the leaves only as a base on which to grow this food. Like industrious farmers, they carefully tend the fungus gardens in their underground chambers, manuring them with their feces and weeding the gardens when necessary. When a queen ant establishes a new nest, she even brings along a pellet of living fungus fragments which will be used to establish new gardens.

## *Tax collectors*

Quite a different class of consumers in the jungle or any other ecosystem are the *parasites,* organisms that take their energy from other living plants or animals—their *hosts*—without as a rule killing them. They are, in a sense, tax collectors who live by taking a small daily toll of the host's energy.

Possibly the most dreaded of all jungle parasites are vampire bats. These notorious beasts, unable to digest any more solid food, live on the blood of other animals. But contrary to popular notions, they do not bite their victims and suck the blood, for their attack is far more subtle. These night-flying bats generally approach their victims while they sleep. Equipped with slightly protruding razorsharp front teeth, the vampires slash the surface of the host's skin and then lap up the blood that oozes from the wound. The attack is so gentle, in fact, that the hosts—whether cattle, hogs, or even men—usually are unaware that they are being victimized. And as is true of parasites generally, vampire bats seldom kill their hosts. Only when the same individual is attacked repeatedly is there any possibility of death due to loss of blood. On the other hand, the bats may kill their hosts indirectly, for they sometimes transmit fatal diseases such as rabies.

But if vampire bats are the most spectacular of jungle parasites, they are far from being the most abundant. Almost

**A tropical caterpillar, victimized by a tiny parasitic wasp, forms a walking dormitory for wasp cocoons. The adult wasp laid its eggs in the living caterpillar's body. Upon hatching, the larvae ate their way through the caterpillar's skin and spun the cocoons from which adult wasps eventually will emerge.**

every kind of animal is subject to the attacks of organisms such as ticks, fleas, mites, lice, and leeches that live on the outside of its body, as well as tapeworms, thread worms, and many other creatures that live inside the intestines and respiratory passages. In addition, vast populations of protozoans, bacteria, viruses, and other microorganisms live inside the tissues, blood, and other body fluids of most animals, taking their daily tax of the host's energy.

So successful is the parasitic mode of life that there are even parasites that live on other parasites. And some of these "hyperparasites" are victimized, in turn, by still other parasites. Thus, there are food chains among parasites just as there are among predators and herbivores. But with parasitic food chains, the animals at each level are smaller, though more numerous, than at the previous level, instead of being larger and fewer in numbers with each link as they are in predator food chains. If this were not so and a parasite were to take more than a small percentage of its host's food, it would kill the host and die itself as a consequence.

### *Plants have parasites too*

The jungle also has many plant parasites. Fungi are by far the most important. But unlike the species that attack potatoes, wheat, and other crops, the parasitic fungi of the

**Cactuslike flowering heads are the only parts of *Thonningia* that ever reach daylight. The rest of this parasitic plant of African jungles remains underground, where it takes its nourishment from the roots of trees. The withered flowering head at the left was probably scorched in a fire.**

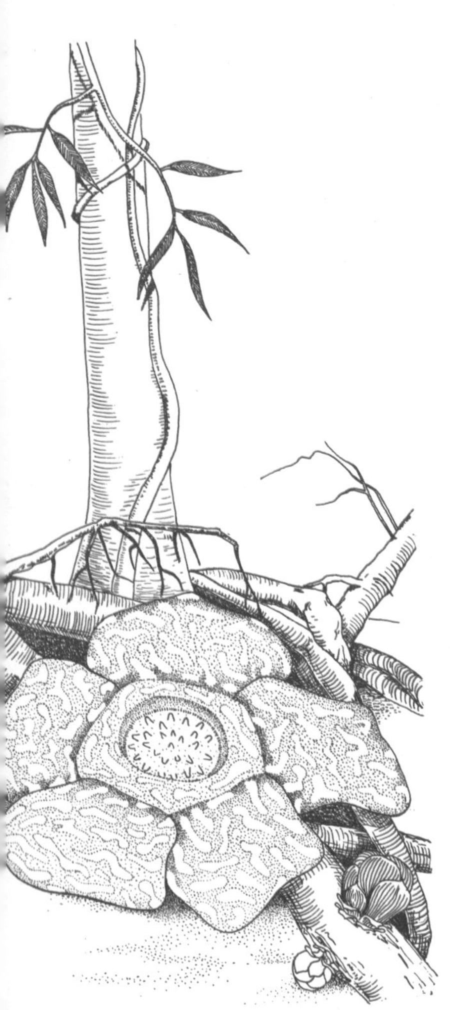

The largest blossoms in the world are produced by species of *Rafflesia*, spectacular parasitic plants of Asian jungles: in one species, the flowers are three feet in diameter. Even the buds are impressive. Growing through the bark of lianas, they develop over the course of nine months into globular masses about the size and shape of large cabbages before finally bursting into bloom.

jungle generally do not seem to cause epidemic diseases. Most hosts apparently have long since become adapted to the presence of their parasites, so that each lives without disastrous consequences to the other.

Far more conspicuous than these microscopic fungi are the many kinds of flowering plants that, like the American dodders and mistletoes, have adopted a parasitic or semiparasitic mode of life. All these plants lack either chlorophyll or some other part of the full equipment necessary for feeding themselves and rely instead on other plants for part or all of their nutrition.

*Thonningia* of the African jungle, for instance, has no chlorophyll. The plant, in fact, spends most of its life underground where it attacks the roots of trees. Only when it is ready to bloom do its brilliant red cactuslike flowering heads finally push through the litter to the surface.

The most famous of all plant parasites is *Rafflesia* of the Malayan jungles, for *Rafflesia* has the largest blossoms of any flowering plant. In one species the flowers are three feet in diameter. But like *Thonningia,* it goes unnoticed for most of its life. Until it is ready to blossom, *Rafflesia* lives as a network of thin branching threads within the stems of large woody vines.

Mistletoes, on the other hand, are parasites of a different kind. These plants, which are common in the tropics as well as in temperate climates, grow mostly high up in the canopy attached to the branches of trees. In tropical Asia and Africa, many species bear strikingly beautiful red or orange flowers. Like most epiphytes, they have green leaves, and with their chlorophyll they synthesize most if not all the organic food they need. But unlike other epiphytes, which depend on other plants only for support, the mistletoes have no underground roots. Instead they depend entirely on the trees to which they are attached for all their water and mineral nutrients.

The part that parasites play in the economy of the jungle, or any other ecosystem, is difficult to assess, but probably not very great. By taking a portion of the host's food and energy for themselves, they undoubtedly lessen its resources for growth and reproduction to some extent. But everything they remove returns into general circulation when the parasites, or the hosts with their parasites, die.

## *Worlds in miniature*

Food chains are found in all ecosystems. The links between jaguars and their prey are an obvious example, but in the jungle food chains exist even in unsuspected places. The tanks of epiphytic bromeliads, for instance, may contain as much as five quarts of water, forming what are, in effect, tiny ponds perched high in the treetops. Miniature habitats of this sort within a larger habitat are known by ecologists as *microhabitats.*

Anyone who looks closely will discover that the microhabitat of a bromeliad tank bustles with activity. Like the jungle itself, it includes producers, herbivores, predators, and decomposers. Occasionally the tanks contain water plants such as mosses and bladderworts, but the major producers are simple green algae and microscopic diatoms similar to those found in ordinary pond water.

The community depends to a large extent on the photosynthesis of these green plants. One researcher studying bromeliads in Jamaica found many animals in the tanks of plants growing in sunny situations where algae were plentiful in the water, but relatively few in the tanks of plants growing in shady places where there was less algae in the water. In addition, the food chains in all bromeliads depend heavily on the dead leaves, drowned insects, and other organic matter that fall into the tanks. In shady places, the food cycles in bromeliads depend mainly on debris of this sort.

Animals in the tank community range from microscopic protozoans, rotifers, and minute water fleas to larger creatures such as worms, water beetles, and even crabs. One species of crab, in fact, is known only from bromeliad tanks. But the most abundant animals usually are the larvae and nymphs of mosquitoes, chironomid flies, dragonflies, and other insects. Several kinds of frogs also are found in the water, and some species even lay their eggs in bromeliad tanks. No one knows for certain how the tadpoles manage to find enough food in the tanks, for most tadpoles are herbivores. It is believed that tadpoles of bromeliad-inhabiting species may be cannibals, living on surplus eggs of their own kind, since females always lay more eggs than can possibly survive.

Of course, the bromeliad tank community is not entirely

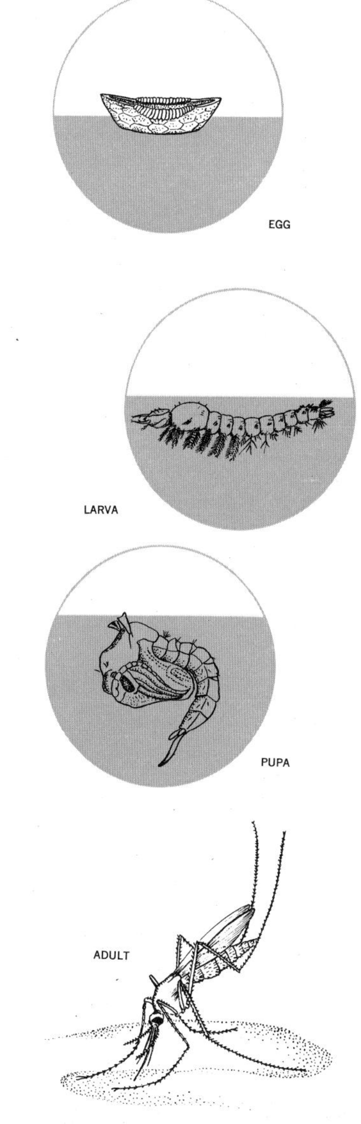


Common—and dangerous—inhabitants of the water tanks of epiphytic bromeliads are *Anopheles* mosquitoes, the carriers of malaria. The eggs, larvae, and pupae all live in water, while the adults are able to fly about and spread the disease. Because the mosquitoes often breed in bromeliad tanks and other inaccessible places, complete eradication of this dreaded jungle disease is almost impossible.

self-contained. It depends in part on food that falls in from outside. And the frogs and insects usually leave the tanks when they mature and find their food elsewhere. But while they live in the tank, their lives are closely linked with those of all the other inhabitants. The plants produce food; some of the animals eat the plants or decomposing organic matter; and still other animals prey on the herbivores and scavengers.

### *Birds, bees, and flowers*

In every ecosystem, the basic relationship between plants and animals is that of producer to consumer. The strands that bind all living things into a complex web of life, however, are not only a matter of food and feeding. Over the course of time, many kinds of plants and animals have developed adaptations that link their lives in many other ways.

A basic need of all flowering plants, for example, is the *pollination* of their blossoms, whether by insects, wind, or some other means. If the plant is to produce seeds successfully, the pollen must somehow get to the stigma of the same or another individual of the same plant species. In the jungle as elsewhere, the commonest pollinators are insects such as bees, butterflies, moths, beetles, and flies. When a jungle tree bursts into bloom, the flowers often attract so many bees that their buzzing can be heard hundreds of feet away. In addition, many tropical flowers are pollinated by hummingbirds, sunbirds, honeycreepers, and a few other specialized types of birds. Even bats are important for the pollination of certain jungle trees.

It is interesting to note, however, that relatively few tropical plants are pollinated by the wind, while in the temperate zone many trees, such as pines, oaks, and cottonwoods, are wind-pollinated. This ensures pollination in the cooler weather of early spring when relatively few insects are active. In the jungle, on the other hand, insects are always plentiful. Moreover, the air usually is very still. And trees of a partic-

**A cutaway section of an epiphytic bromeliad reveals the series of small reservoirs formed by the bases of the plant's leaves. In addition to the snail exploring this specimen, algae, frogs, insects, worms, protozoans, and many other forms of life may inhabit the tiny treetop ponds formed by bromeliads.**

ular species generally are widely scattered in the jungle, so that the trees would have to produce enormous quantities of pollen if they depended on the wind to carry it to other trees of the same species.

The relationships between flowers and their pollinators sometimes are highly specialized. Hundreds of kinds of figs grow as trees, vines, and stranglers in the tropics. In every case, their flowers are pollinated by very small wasps that lay their eggs in the young flowers and spend almost all of their lives inside the developing fruits. The adjustment of plant and insect to each other is so precise, in fact, that it is believed that every species of fig has its own special kind of fig wasp.

### *Attracting pollinators*

Animals, of course, do not pollinate flowers intentionally; they come in search of nectar and pollen for food. But in the process of taking this food, some of the pollen usually adheres to their bodies. When they visit another plant, the pollen rubs off on the female portion of the second flower and fertilizes the ovules which will develop into seeds.

In the course of evolution, many flowers and their pollinators have developed delicately adjusted coadaptations that make successful pollination possible. The production of nectar, for instance, is probably of no direct value to the plant; it is useful only because nectar attracts pollinators. Sometimes this device is carried to extremes. The scarlet *Erythrina* of the West Indies is often called the "cry-baby tree" because so much nectar drips from its flowers.

Yet different kinds of flowers are adapted in an almost incredible variety of ways for pollination by different animals such as birds, bees, or moths. Many long-tubed, trumpetlike flowers, for example, are pollinated only by butterflies and moths. The mouth parts of these insects form long tubes that remain coiled like watchsprings when not in use. When the moth or butterfly visits a flower, the tube is uncoiled so that it can probe for nectar hidden deep at the base of the trumpetlike blossom.

There are variations even on this particular pollinator-flower relationship, however. Large night-flying hawk moths

While collecting pollen with which it will provision the nest where it lays its eggs, a carpenter bee inadvertently assists in the pollination of a melastome flower. Some of the pollen adhering to the bee's fuzzy body when it visits one blossom is likely to rub off on the next flower it visits, thus assuring cross-fertilization of the plant species.

**With its tubular mouthparts unfurled to form a drinking straw, a glass-winged butterfly sips nectar from the tiny florets of a plant in a Peruvian forest. Any pollen that adheres to its legs and abdomen will be carried to another flower.**

are very common in the jungle, and they mostly visit flowers that open fully only at night. The flowers they visit, moreover, tend to be white or very pale in color and usually have a strong, sickly sweet scent. The light color and strong odor make them easy to locate in the dark. The blossoms visited by day-flying butterflies, in contrast, tend to be brightly colored and less powerfully scented.

Birds are quite different from insects in their reactions to flowers. Birds are not very responsive to odors, and their color vision is quite unlike that of insects. While bees are most strongly attracted to blues and yellows, birds are said to prefer red, orange, blue, yellow, white, and green, in that order. Hummingbirds seem definitely to prefer bright scarlet to any other color. And, as might be expected, there are many scentless, bright red flowers in the South American

**Hovering on rapidly beating wings, a hermit hummingbird uses its long bill and even longer tongue to probe for nectar hidden in the depths of a tubular blossom. Like butterflies, nectar-feeding hummingbirds, sunbirds, and honeycreepers are important pollinators of many jungle plants.**

jungles and in other areas such as California where hummingbirds are common.

Bees, furthermore, usually alight on the flowers they visit, while hummingbirds hover in front of the blossom as they feed. Thus it is not surprising that the bird-pollinated flowers of many tropical American vines and trees project well beyond the surrounding leaves, either on long upright spikes or on flexible stems that sometimes dangle down for several feet. And they usually lack the alighting platforms provided by the petals of many bee-pollinated flowers.

Bats also play a large role as pollinators in the tropics. Until recently, their importance was underrated, since they fly by night or at dusk when it is difficult to watch them at work. Although the bats of temperate climates feed mainly on insects, many tropical species visit flowers and live entirely on fruit and nectar. When they feed, these bats usually alight and grasp the flowers with their claws. The following morning the claw marks remain as visible evidence of their visits.

Like hummingbirds, they need a clear run-in, and bat-flowers usually are in exposed positions. Sometimes they hang from the trees on long stalks, but they may also be cauliflorous blossoms borne on the bare trunks of trees below the branches. Whatever families they belong to, bat-pollinated flowers have various features in common. Typically they are large, they open at night, and usually they are dull whitish, greenish, or brownish in color with strong, disagreeable odors. *Oroxylon*, a bat-pollinated tree of Malaya, for instance, is sometimes called "the midnight horror"; its flowers open about ten o'clock in the evening and by midnight begin to emit a horrible stench.

**The cannonball tree of the American tropics, a relative of the Brazil nut tree, is an example of a bat-pollinated species. The cauliflorous blossoms are borne on short stems that grow directly from the lower part of the trunk, where bats have plenty of room to fly in to the flowers.**

### *Sowing the seeds*

In addition to pollinating their flowers, animals also benefit plants by distributing their seeds. Sometimes they transport seeds by carrying burrs or hooked fruits in their fur. But, except along roads and trails, fruits of this sort are rare in

the jungle. More commonly, jungle birds and mammals scatter the seeds when they eat the fruits. Monkeys, for example, are well known for their habit of plucking and then dropping far more fruit than they eat. Seeds that are eaten often pass unharmed through animals' digestive tracts and are thus transported for great distances.

Fruits of many kinds are very common in the jungle and form an important part of the diets of squirrels, monkeys, parrots, peccaries, civets, fruit bats, and other vegetarians and mixed feeders. Even carnivores such as tigers and jaguars seem to enjoy eating fruit from time to time. And in the case of many epiphytes, the ants that often live among the roots frequently carry their seeds from place to place.

Many fruit eaters have become especially equipped for their diet by developing structures such as the extraordinary fruit-plucking and nut-cracking bills of parrots, toucans, and hornbills. Plants likewise have evolved fruits with colors, odors, and flowers attractive to animals, and have developed other characteristics such as seeds that are resistant to digestive juices. The preference of birds for bright reds, for instance, is certainly responsible for the abundance of scarlet fruits and fruit stalks in the jungle.

Animals are not the only means by which the seeds of jungle plants are distributed. Many trees and vines have

**Individual blossoms of the cannonball tree are five inches across and sufficiently sturdy to support visiting bats in search of food. Following pollination, the blossoms mature into large, hard-shelled, seed-filled fruits that resemble rusty cannonballs.**

lightweight fruits or seeds that are equipped with wings or parachutes. These fruits and seeds are transported by the wind. Others, in contrast, have very heavy fruits that crash straight down through the foliage to the ground.

To some extent, the means of seed distribution depend on the kind of growing conditions required by the seedlings. Trees characteristic of gaps and clearings, such as the tropical American cecropias and vismias, as well as most vines, need a great deal of light when they are young. But gaps and clearings in the jungle normally are far apart, and these plants usually have seeds that can quickly be carried some distance from the parent tree by birds, bats, or the wind.

The seedlings of trees characteristic of undisturbed jungle, on the other hand, usually grow best in deep shade beneath an unbroken canopy. Their seeds are generally adapted for dispersal only over short distances. In the Guyana jungle, the whitish flowers of the greenheart tree are succeeded by heavy, egg-shaped fruits that weigh several ounces. As they drop from the tree—often in huge numbers—many remain where they land while others are eaten by animals and a few are carried undamaged for short distances. But most of the

**Flying foxes, the large fruit-eating bats of Malaya and the East Indies, are some of the many animals that take advantage of the jungle's abundant supply of fruit. Figs, mangoes, papayas, breadfruit, bananas, and even coconuts are some of their favorite food sources.**

fruits are scattered within a radius of only one hundred yards or so of the parent tree.

Unlike the seeds of many temperate-region plants, those of the greenheart cannot lie dormant in the soil for months or years before germinating. Soon after they fall, the undamaged seeds begin to sprout and within a few months the area is covered with a thicket of saplings two or three feet high. Very few of them become mature trees, however. Only one or two live long enough to grow up to the light when a break finally appears in the canopy.

Many other shade-tolerant trees also have extremely large, heavy fruits. The Brazil nut, a giant tree of the Amazon jungle, has an enormously heavy fruit resembling an earthenware pot which contains the nuts. And a number of African trees have fruits as large and almost as heavy as cannonballs. Their great weight enables them to crash down through the layers of twigs and leaves between the treetops and the ground. Their size also is advantageous in another way, for it means that they contain large stores of food on which the young seedlings can survive until they have enough leaves to produce food for themselves.

**Parrots, like toucans, hornbills, macaws, and a number of other birds of the treetops, live primarily on a diet of fruit and nuts. Stout bills and nimble feet serve as excellent tools for manipulating their food.**

*Dischidia*, a common epiphyte on tree trunks in Malayan jungles, is a myrmecophyte that depends in part on its ant tenants for nourishment. Certain of *Dischidia*'s leaves develop into hollow, vase-shaped containers which are usually inhabited by ants. The plant's roots grow into these containers, apparently taking nourishment from debris carried in by the ants.

### *Ant plants*

The most interesting but perhaps least understood of all plant-animal relationships in the jungle are those that exist between ants and certain kinds of plants. As we have already seen, ants frequently live among the roots of most epiphytes, and in some they live in the debris that collects in bracket- or sac-shaped leaves. In a few, such as the curious swollen-stemmed *Hydnophytum* and *Myrmecodia*, the ants even live in tunnels inside the epiphyte's tissues.

But many jungle trees, treelets, and vines also are invariably inhabited by ants. On *Tachygalia myrmecophila*, a very tall tree of the Amazon forest, ants are always found in the hollow spaces inside the tree's swollen leaf stalks. *Triplaris*, another large South American tree, also harbors ants within its hollow twigs. And in one kind of climbing palm, or rattan, of Malaya, ants live in the hollow sheaths of the leaves.

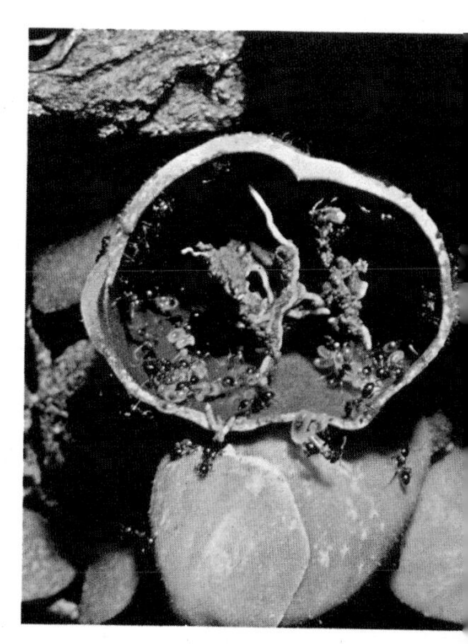

**With its top cut off, a vaselike leaf of *Dischidia* is seen to be a hollow container sheltering a colony of ants, complete with eggs, larvae, pupae, and bustling adults. At the center are several roots encrusted with humus formed from the debris brought in by the ants.**

Plants that are invariably inhabited by ants are called *myrmecophytes*, or ant plants. They exist only in the tropics, though why this should be so is not quite clear. The ants take up residence when the plants are young and are found in every individual of the host species. Even more extraordinary is the fact that each kind of host plant generally shelters its own species of ant, and the ants usually live nowhere but on their host plant.

In some—probably most—instances, these plant-inhabiting ants, like many free-living species, live in association with a population of aphids or mealy bugs. The ants care for these little insects much as a farmer tends his cattle, for the aphids and mealy bugs suck the plant juices and secrete a sugary liquid, honey-dew, on which the ants seem to depend to a large extent for their food. In addition, on many ant plants, tiny glandular outgrowths containing fat and protein form on the leaves and elsewhere; these food bodies, as they are called, are bitten off and eaten by the ants.

But what exactly is the relationship between the ants and the plants they inhabit? The ants usually receive free board and lodging. But does the plant receive any benefit in return? In most cases we are not sure, though the existence of these curious ant-plant associations has been known for over a hundred years.

**Excavating a new home, a queen ant bores an opening into one of the characteristically paired thorns of a Central American bullhorn acacia. Later she will hollow out the thorn and lay eggs inside. Normally every thorn on a bullhorn acacia is inhabited by ants.**

## *Ants in the acacias*

When myrmecophytes were first studied, it was thought that the ants protected the plants against attacks by other insects. The English naturalist, Thomas Belt, who studied the ants of the bullhorn acacias of Central America, suggested that the ants formed a sort of standing army that drove off both leaf-cutting ants and mammals that might be tempted to browse on the acacia leaves. For many years his notions were regarded as a romantic myth, but more recent studies suggest that the partnership between plant and ant may be even more remarkable than Belt supposed.

Bullhorn acacias are small trees of open dry forests and thickets, rather than of the tall jungle. The ants live inside the very large paired thorns that give the trees their name. They feed on sugar produced by nectar glands on the leaf stalks and on small oil- and protein-filled growths found at the tips of leaflets.

Daniel H. Janzen, who recently studied bullhorn acacias in Mexico, found that the ants not only drive off invading insects and larger herbivores as Belt had supposed; they also protect the trees from vines and other plants that might smother them. The ants do this by biting off the growing tips and tendrils of any plant other than their own acacia growing in the neighborhood. As a result, the trees are surrounded by open spaces, allowing them to grow freely. The acacias, in fact, are so dependent on their ants that, if they are treated with insecticides to kill the ants, they do not flourish. Before long, the trees are overcome by competition from surrounding vegetation.

It is clear that the bullhorn acacias and their ants live in a very highly developed partnership, and there are probably other cases where the situation may be mutually beneficial. The hollow stems of *Barteria fistulosa*, a small tree in Nigeria, are always inhabited by fiercely biting black ants. The jungle people claim that no other tree will grow within a certain distance of *Barteria*. It may well be that the ants prune surrounding vegetation in much the same way as do the ants of the Central American acacias. But it does not follow that the ant-plant relationship is the same in all myrmecophytes. In some cases, the ants may be merely guests that contribute little or nothing in return for their lodging.

**An important food source for the ants associated with bullhorn acacias are the small but nutritious beadlike growths at the tips of young leaflets (*top*). The ants also feed on sugar secreted by craterlike nectar glands on the leaf stalks (*bottom*).**

## *The web of life*

All in all, the jungle is a world of plants and animals whose lives are intertwined in countless ways. The three-toed sloth provides an especially good example of the subtlety of some of these relationships. As we have already seen, the sloth feeds mainly on leaves of the cecropia tree. The sloth in turn is commonly greenish in color due to particular kinds of algae that grow only on its hair. But the great chain of life does not end here, for a certain kind of moth is often found among the sloth's hair, where it lays its eggs. And the moth's larvae in turn feed exclusively on the algae that grow only on the sloth's hair.

There are many other ways in which the lives of jungle

plants and animals are interrelated. Some, as we have seen, provide others with food; some compete with others for food, light, or other necessities; and some depend on others for such varied needs as shelter, protection from enemies, and, in the case of plants, pollination and seed dispersal.

Yet the system, complex as it is, seems to have built-in controls that keep it in balance. Herbivores, for instance, eat plants but seldom destroy them, and carnivores manage to find enough food without exterminating their prey species.

This balance results in part because the jungle has existed for millions of years. Over the course of time, the plants and animals have become increasingly adapted to each other. As the herbivores have become more efficient at eating plants, the plants have evolved defenses against the herbivores. Some are armed with stout thorns, and others seem to contain natural insect-repellent substances that protect them against caterpillars and other plant feeders.

The world of the jungle, in short, is one where all the elements of the ecosystem exist in a kind of harmony. Yet how the balance is maintained we cannot yet say. Certainly there are strands that bind all living things together into a web of life, but it will take a long time to decipher completely the pattern they form.

But we do know much about what happens when the balance is upset. Man's interference, in particular, can have effects that are often catastrophic, and these we must now explore.

**Secure on its aerial perch, a tamarin surveys the green realm of the South American rain forest. Like all jungle creatures, the little monkey inhabits a world filled with danger. Yet as long as its jungle home survives, the species can live in harmony with its environment and continue to thrive.**

# *Man and the Jungle*

The jungle is not an easy place for man to live in. Until well into the twentieth century, the humid tropics remained one of the world's most thinly populated regions. Over vast areas in South America, Central America, and tropical Africa, the life of the jungle went on without interference, just as it had during the millions of years before man evolved. Less than one hundred years ago, even the Malay Peninsula, now one of the most highly developed areas in the tropics, was covered with an almost unbroken forest. Along the coasts and larger rivers there were towns and villages built by colonists from Sumatra, India, China, Europe, and elsewhere. But the only people who lived in the forest itself were small groups of primitive Sakai nomads.

Now all this is changing. Today a visitor traveling by plane will find few regions where the virgin jungle stretches unbroken from horizon to horizon. In most places, he will see a mixture of farms, villages, plantations, and airstrips, linked together by miles of new roads. Here and there in the vivid green of tropical vegetation he will see great gashes of exposed red soil, the work of bulldozers and subsequent erosion by heavy rains.

This sort of modern development is very recent and still limited in extent. Much more jungle has been destroyed by the native farmers' age-old methods of cultivation. Corn, cassava, and other food crops are planted in small clearings in the jungle. But cultivation methods are primitive, and, for reasons we will soon examine in more detail, crops can seldom be grown on the same plot two years in a row. Instead, the land is abandoned and a new patch of jungle is felled each year. As a result of this so-called *shifting cultivation*, thousands of square miles of jungle have already been transformed into patch-works of tiny fields, second-growth forests, and wasteland. And now, because of rapidly expanding populations, the process is being accelerated. Each year more and more of the virgin rain forest is being nibbled away by shifting cultivation.

### *Punans and pygmies*

Men have not always been destroyers of the land on which they live. For thousands of years, primitive people have inhabited jungles in many parts of the world. But since they had no knowledge of agriculture and used only stone or wooden tools, their influence on their surroundings remained insignificant. Even today a few surviving jungle tribes continue to live in harmony with their environment.

One such tribe is the Punans, a shy, light-skinned people who live in the forests of Borneo. In recent years the Sarawak government has persuaded many of them to settle in villages. But others still live as wanderers in the jungle, just as their ancestors did for uncounted centuries.

The Punans have no permanent houses. They build flimsy shelters of branches and palm leaves but use them for only a month or two before moving on to new locations. Instead of growing crops, they depend mostly on wild sago extracted from the stems of a palm that grows in swampy parts of the forest. They also hunt for wild pigs, monkeys, and any other game they find.

**Pygmies, nomadic hunters and food gatherers of the Congo rain forest, build simple waterproof huts of saplings thatched with large overlapping leaves. Campsites are occupied only as long as game and wild food plants last in an area. Then the huts are abandoned and the tribe moves on to new territories.**

Although most Punans now possess shotguns, until a few years ago they hunted mostly with blowpipes. These weapons, made by boring a tube through a thin hardwood stem, were used for shooting poisoned darts. Even their poisons were products of the jungle; they used both the juice of the ipoh tree and that of a species of *Strychnos*, a plant which is related to the vine that provides the Indians of South America with the main constituent of their famous curare arrow poison.

The Punans cannot work metals. They obtain their knives, guns, tobacco, and other things they cannot find or produce themselves by bartering with their more advanced neighbors, such as the Kayans, Keniahs, and Ibans. These people live in villages, and are always eager to obtain the dammar gum, ilipe nuts, and other products that the Punans collect in the jungle.

Their life is necessarily a wandering one, however. Many useful things can be found in the jungle—timber, roofing materials, quite serviceable string, and even bark which can be made into clothing. But food for humans is scarce. Edible fruits, nuts, and other plant foods are hard to find, and the supply of animal food is highly precarious, even for hunters equipped with traps and modern firearms. Once the small amount of game and wild sago in one place has been exhausted, the small family groups of Punans must move on to new areas.

Their way of life is very similar to that of other simple jungle folk. In the eastern Congo in Africa, the pygmies also live mainly by hunting animals with bows and poisoned arrows and, like the Punans, barter jungle produce with settled tribes who farm the forest clearings. Some of the most primitive South American Indians also live mainly as hunters and food gatherers. But today, people who depend entirely on wild plants and animals scarcely exist outside the tropical jungles. And wherever they live their numbers, like those of other predators, can never be very large. These nomadic food gatherers must always remain a rather unimportant part of the ecosystem.

**Armed with bows and poison-tipped arrows, Congo pygmies prepare for the hunt. They also capture game in nets woven from vines. To supplement their diet, the Pygmies gather forest fare such as nuts and mushrooms, and barter forest produce with farming tribes for rice, manioc, and other crops.**

## THE FOOD GATHERERS

Unlike African Pygmies and the Punans of Borneo, relatively few jungle people live entirely on what they can hunt and gather in the wild. Most of them live in more or less permanent settlements and cultivate food crops in clearings. Yet even these jungle farmers supplement their diets with the bounty of the surrounding forest.

Through long experience they have learned which plants are poisonous and which have medicinal uses; which must be consumed as soon as gathered and which may be stored. From forest plants they collect edible leaves, fruits, nuts, roots, seeds, and mushrooms and other fungi. They also gather grasshoppers, termites, grubs, frogs, lizards, fish, and mammals of all kinds. And they often use wild honey as a source of sugar.

At the left, a southeast Asian youth returns from a foraging expedition with a harvest of *Parkia* pods filled with seeds that are cooked for food. On the right, two Malays have collected a supply of jackfruit, which can be eaten fresh, cooked as a vegetable, or preserved in syrup.

## *Jungle farmers*

Most jungle people today live as farmers. Although they usually take advantage of any available game, fish, or wild plants, most of their food comes from cultivated crops grown in temporary forest clearings. Trees are felled and burned, and crops are planted. But since manures and fertilizers are seldom used, the fertility of the soil quickly decreases. Within a year or two the soil will support no more crops; the field is abandoned, and a new clearing made.

Obviously this method of shifting cultivation results in the destruction of huge areas of the forest. Yet it is still the commonest way of raising crops in the jungle. It is said that over two hundred million people in South and Central America, Asia, and Africa depend on this system for their staple food. With only minor variations in detail, it is used by people as different as modern Brazilians, many African tribes, and the Malays and Dyaks of Borneo. Even the ancient Mayas of Mexico farmed in a similar way, as did the English colonists in Virginia in the seventeenth century. For people with limited resources at their disposal, it has always seemed to be the best way of producing food in a forest environment.

**In a forest clearing in northern Borneo, the clustered huts of jungle farmers are surrounded by a crop of manioc, planted at random among the fallen trunks of jungle trees. Throughout the tropics, this useful plant, also called cassava, is the crop most commonly planted by the millions of farmers who practice shifting cultivation.**

The first stage in the cycle of shifting cultivation, also known as slash-and-burn agriculture, is to make a clearing in the jungle. Some of the fallen trees may be used for fuel or to construct houses, but most are simply piled up and burned.

But what effect does shifting cultivation have on the plants and animals of the jungle? Can the forest ever return to land that has been cultivated? And why does soil that has supported luxuriant forest for thousands of years so rapidly become sterile and useless once the land is cleared? To answer these questions, we must follow the changes that take place when an area of jungle is felled to make room for crops.

### *Clearing the land*

Sometimes it is one farmer and his family who make the clearing; sometimes a whole village works together. However the job is done, clearing enough land to support even one family involves a great deal of work. First the undergrowth and vines are cleared with machetes, then the larger trees are felled. Though the work takes a long time, it is amazing



**Ancient trees go up in smoke as Brazilian farmers make a clearing in the jungle.**

in Borneo to see how quickly two men can cut through the trunk of a giant tree using only biliongs, diminutive axes less than a foot long.

Some of the biggest trees are usually spared, either because the labor of felling them is not worthwhile, or because of superstitions that the spirits of the trees will bring bad luck if they are disturbed. But the rest of the trunks and branches are piled up to get as dry as possible in the brief rainless periods that occur even in the humid tropics. And then the trees and brush are burned. If necessary, the half-burned wood is gathered up a second time and burned again.

In different countries the temporary fields in jungle clearings have different names. In Mexico they are called "milpas," in the Philippines, "kaingins," and in Malaya and Borneo, "ladangs." But everywhere they have the same confused, untidy appearance. Even after a second burning the fields remain littered with charred logs and half-burned brush. Scattered everywhere are standing tree trunks, most of them dead and blackened, and the stumps of buttressed trees cut off six or seven feet above ground.

**In an unweeded clearing in South America, young banana plants grow up around the buttressed stump of a fallen forest giant. Soon after the first crop has been harvested, the clearing probably will be abandoned. Behind the unburned log, a thicket of young cecropia trees already has sprung up on ground that was cultivated a year or two earlier.**

## *Planting the crop*

The farmer makes no attempt to plow his land. Using only the simplest of tools—often just a digging stick but sometimes a metal hoe—he plants his seeds directly in the two inches or so of topsoil, a layer rich in humus and wood ashes. He does not even plant the crop in regular rows but is content instead to let the plants struggle up as best they can among the charred logs and brush.

The crops may be of one or several different kinds. The Hanunóo people of the Philippines are said to cultivate as many as forty of fifty different kinds of plants in their fields. But throughout the tropics, one of the commonest food crops is cassava, or manioc. The large starchy roots of this bush are used for making both tapioca and cassava meal, a kind of flour. In southeast Asia and some other areas, the farmers generally grow hill padi, a variety of rice that does not need irrigation. Other important crops are corn, bananas, sweet potatoes, and yams, while pineapples and many other plants also are grown in smaller quantities.

**In a more permanent type of agriculture, a clearing in the Philippine jungle has been planted with a variety of carefully tended crops. Beneath a scattering of coconut palms, papayas laden with fruit are intermingled with several neatly cultivated rows of pineapples. In the foreground are the bannerlike leaves of a banana plant.**

## THE STAPLE FOODS

Hundreds of different food crops are cultivated by jungle farmers. But certain plants, because they are easily grown, nourishing, and versatile in their uses, are favored as the staples of shifting-cultivation agriculture. Among the grains, some of the most important are rice, millet, and corn. Major root crops are manioc, sweet potatoes, and taro, or coco-yams. Beans, breadfruit, bananas, plantains, papayas, oranges, mangoes, avocados, squash, and melons also are planted.

On the left is a field of manioc, also called cassava in South America. The leaves are sometimes eaten as greens, but manioc is grown mainly for the bunches of five to ten sausage-shaped underground tubers that grow at the base of each plant. The two-foot-long tubers are filled with starch and may be boiled, fried, or pounded into a flour that is used in many ways. Our familiar tapioca, used for pudding, also is made from manioc.

On the opposite page (*left*) fingerlike young bananas grow in rings around a flower stalk. Another popular fruit in the tropics is the papaya (*opposite page, right*).

Breadfruit, which belongs to the mulberry family, is a native of Pacific islands that is grown throughout the tropics. The eight- to twelve-inch-long fruits, which are unusually rich in carbohydrates, are eaten raw, boiled, baked, or ground into flour that can be stored for later use.

Pineapples, now cultivated in many tropical countries, originated in South America, where several varieties still grow wild in the jungle. Although they grow on the ground, pineapples are part of the bromeliad family, whose members are the most conspicuous of all epiphytes in the New World tropics.

As soon as the crop begins to grow, many other plants spring up and compete for space and light. Some are sprouts from the roots of trees and vines that were not completely killed by the fire. Many others grow from seeds that are continually carried in by the wind or by birds and other animals. And some grow from seeds that have been lying dormant in the soil; unlike those of forest trees, the seeds of many plants that flourish in clearings remain viable in the soil for long periods, ready to germinate as soon as the forest cover is removed.

At first the farmer uses his machete to slash away the tangle of weeds and vines that threatens to overrun his crop. But once the crop has been harvested, he lets this rampant growth go on unchecked and abandons the clearing. Experience has taught him that it is generally useless to attempt to cultivate the field again. The yield on the second crop would be too poor to merit the labor of cultivating it.

### *Worn-out soil*

To understand why this is so, we must take a look at what happens to the trees, and especially to their mineral constituents, when the forest is felled and burned.

The stems, roots, and leaves of a living tree or any other plant contain a large percentage of water. But, as we have already seen, they also contain compounds of carbon and nitrogen. Other elements necessary for plant growth, such as phosphorus, potassium, calcium, magnesium, iron, and sulfur, also are present in lesser amounts. While the tree is alive and growing, it is constantly accumulating carbon from the carbon dioxide taken in from the air during photosynthesis. At the same time, the roots are taking up varying amounts of the other elements as compounds dissolved in soil water.

When the jungle is undisturbed, the trees, besides accumulating minerals, are also constantly losing large quantities of the same elements in the form of falling leaves, branches, and so on. On the ground, bacteria and other decomposers release the elements from the dead plant material, but the roots take them up again so quickly that very little can be washed away to streams by drainage water. The mineral cycle in the jungle, in short, is almost a closed one, and most

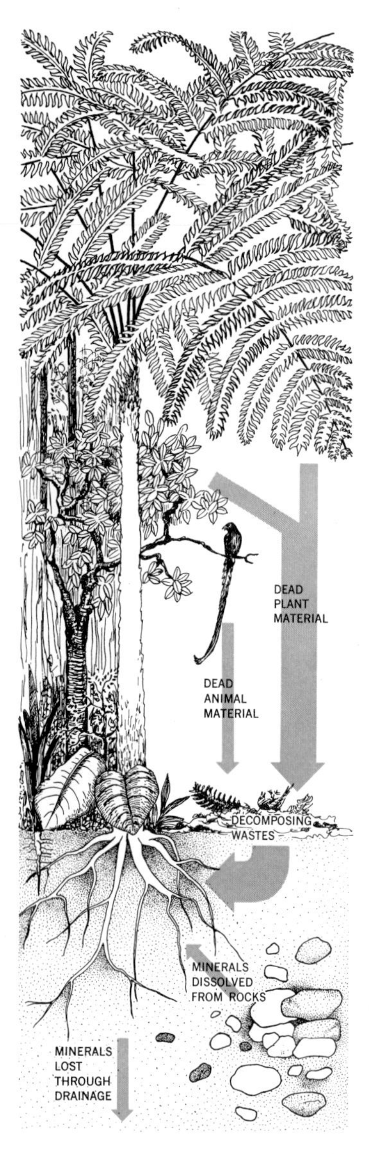


In undisturbed jungle, the vital mineral nutrients circulate in an almost closed cycle. All the minerals used by plants eventually return to the soil where they can be reused, and the small amount of minerals lost in drainage of water is balanced by the continuing release of minerals dissolved from rock particles. Widths of arrow indicate the relative amounts of minerals passing through various stages in the cycle.

of the minerals originally derived from the underlying rock are stored in living plants and in the litter on the ground rather than in the soil.

When the jungle is felled and burned, this neat, self-sustaining cycle of minerals from the soil, to living plants, and back again to the soil is completely shattered. Almost all the nitrogen and sulfur and a great deal of the carbon disappear into the atmosphere in the form of gases and smoke. The remaining elements are converted into simple inorganic compounds in the wood ash. Since most of these compounds are fairly soluble in water, great quantities are washed away by the first heavy rainstorm after the fire.

Even so, quite a high concentration of mineral nutrients remains available for a while in the surface soil. But once the first crop and accompanying weeds have grown, the supply drops sharply. Harvesting the crop means an even further loss of nutrients. Ten thousand pounds of cassava tubers, the average yield for a quarter of an acre, for example, contain about twenty-two pounds of nitrogen, three pounds of phosphorus, and fifty-eight pounds of potassium. As a result, the available mineral content of the soil at the end of the growing season is almost exhausted. As far as the farmer is concerned, the soil is worn out and the plot must be abandoned.

### *The minerals return*

The land is not ruined for all time, however. A clearing, once cultivated, can eventually become productive again. Jungle farmers long ago learned that if they let plants grow wild on the clearing for a number of years, the fertility of the soil eventually would be restored. In time it would be possible to clear the land again and grow another crop.

This is so because as new kinds of plants take over in the clearing the supply of mineral elements gradually improves. Small amounts of nitrogen are added by rain water, and much larger quantities are added by nitrogen-fixing bacteria in the soil. These bacteria are able to convert atmospheric nitrogen into compounds that can be used by plants. In addition, phosphorus, potassium, and other elements are slowly released by the breakdown of rock particles. Further supplies of minerals are brought to the surface from lower layers by the new trees, which frequently have deeper roots than the original forest trees. In time litter begins to cover the soil, and gradually the amount of mineral nutrients stored in the cover of living plants returns to its former level. Now at last the land can be cleared again and new crops grown.

Even so, the slash-and-burn technique of shifting cultivation is a wasteful method, for it requires the use of great areas of land. For every acre under cultivation, another fifteen or twenty acres must lie fallow while the soil slowly recovers its fertility. Yet the millions of hungry people living in jungle regions must be fed. And at present, we still have so much to learn about tropical agriculture that we know of no other practicable way for them to raise their food.

When the jungle is cleared for cultivation, the self-sustaining mineral cycle of undisturbed forest is broken and the available minerals in the soil are rapidly depleted. This results because nutrients are removed from the ecosystem when crops are harvested, and because erosion washes away more of the soluble minerals from cultivated land than from the well protected soil of the jungle.

## *After the fire*

When a farmer makes a clearing in the jungle, much more than the mineral content of the soil is changed. Suddenly and dramatically, a delicately balanced ecosystem is destroyed and an entirely different environment is created.

Beyond the fringes of the clearing, the life of the jungle goes on more or less as usual. But in the clearing itself, whole new communities of plants and animals gradually take over. Once the trees are gone, for example, the monkeys, toucans, tree frogs, flying squirrels, and other canopy dwellers lose both their homes and their livelihoods. Even if the farmer leaves a few tall trees standing, most of these creatures disappear entirely from the clearing.

The same is true at every other level of the jungle. In one stroke, the dim, sunflecked world of the undergrowth vanishes. Instead of a shady place where the air is always moist and the temperature seldom varies, the clearing becomes a bright sunlit place with drier air and fluctuating temperatures. The moist, litter-strewn forest floor, in turn, is replaced by expanses of bare soil that, in the middle of a sunny day, dries out and becomes almost too hot to touch.

After the fire, plenty of logs and dead branches still lie on the ground, but most are charred and have become dry and brittle. The fungi and microorganisms that had lived on the litter and among the decaying logs cannot survive under these new conditions. Most of the small animals either die in the fire or flee into the nearby forest. Yet other creatures thrive under the new conditions. Soon after the fire, the

**Once the crops have been harvested in a jungle clearing, the soil's fertility usually is so badly depleted that the field is allowed to revert to wilderness. Here in a newly abandoned clearing in North Queensland, Australia, the untended earth already has been carpeted with low-growing weeds, ferns, and vines. Within months, the field will be covered with a nearly impenetrable tangle of vines, shrubs, and saplings.**

clearing is likely to be invaded by lizards, tiger beetles, and various kinds of bees and solitary wasps that are rarely seen inside the forest.

Once the farmer's crop and the inevitable weeds begin to grow, life in the clearing becomes much more varied. With a supply of food available, herbivores—especially insects—begin to multiply. And insect-eating birds such as flycatchers, martins, swifts, and swallows come seeking the insects. Studying a recently made clearing in the Guyana jungle, William Beebe long ago noticed that the field soon was inhabited by sixty-five species of birds "which in no sense could be termed jungle birds." The first to arrive was a ground dove that landed one day on a stump to eat the seeds from a pile of weeds left by laborers who made the clearing. Before long, the dove was followed by tanagers, grassfinches, kingfishers, rails, kiskadees, and many others. The same is true in Africa where weaverbirds, for example, often take over any large trees left standing in a clearing. These birds nest together in huge colonies but never occupy trees in the mature forest.

### *The weeds take over*

Really dramatic changes do not begin until the field is abandoned. At first a few remnants of the corn, cassava, or bananas struggle on for a while. But once the farmer stops cultivating his field, the weeds take over in earnest.

Before long the clearing is covered by a maze of herbs, grasses, ferns, bushes, and young trees, few of them species that grow in the interior of the jungle. And lacing it all together are fantastic tangles of vines that loop and twine their way over and around every support. Many of the plants, moreover, are studded with fierce thorns. To walk through such a tangle is impossible, and even to hack a path with a machete is slow, hard work.

Many of the first plants to colonize the clearing are weeds that were present before the crop was harvested. After a few months, however, these early colonizers disappear as other plants, mostly young trees, grow up and begin to overshadow their competitors.

As for animal life, relatively few long-term observations

have been made on the changes that take place in abandoned clearings. Certainly there are not enough detailed records to piece together any kind of a continuous story. But there are at least a few scattered observations that can give a general idea of the kinds of things that happen.

Until trees begin to grow again in the clearing, for instance, we know that the mammal population in the fields and first growth of tangled thickets is restricted mainly to small ground dwellers that are active only at night. In West Africa, cane rats and Old World porcupines are common, and sun squirrels, musk shrews, and giant rats are also found. Until some sort of forest cover returns, most larger mammals probably keep away from the clearings except at night.

Many kinds of lizards, snakes, toads, and frogs also move into the young second growth. But the newcomers usually are quite different species from those that live in the nearby jungle. Although leafeating insects are much less numerous than in the forest, there are many other small plant-feeders. These form the food of a great variety of predatory insects, such as hunting wasps, carabid beetles, and tiger beetles, which are eaten in turn by the lizards, snakes, and other larger predators.

Among the animals that invade abandoned fields in Old World jungles are porcupines, such as the African brush-tailed porcupine. In addition to dense spines on its back, this aptly named rodent has a tuft of bristles at the tip of its nine-inch-long tail. During the day it sleeps in a burrow, but at night it emerges to feed on plant material and insects.

### *Musangas move in*

In Nigeria, where the changes in abandoned clearings have been carefully studied, the second growth of vines, shrubs, and young trees reaches a height of about fifteen feet within three years of abandonment of the field. By this time it looks like a low forest and is almost always dominated by the vigorously growing musanga or parasol tree. This tree, whose large leaves composed of radiating leaflets are somewhat similar to those of the American buckeye, is a characteristic feature of roadsides and clearings everywhere in the African jungle.

The musangas grow so fast that, by the time they are fourteen years old, the trees may be seventy or eighty feet

**The unmistakable pattern of musanga leaves against the sky is a familiar sight in Africa. These fast-growing, short-lived trees, the first to invade abandoned fields in African jungles, flourish wherever the canopy of the original jungle has been disturbed.**

high. By this time, the umbrella-shaped crowns responsible for the name "parasol tree" have formed a closed canopy. The canopy is much more uniform in height than was that of the original forest. And it is practically unbroken by the crowns of any other trees, though saplings of many different species struggle in the shade beneath the musangas.

Perhaps because of their rapid growth, the musangas are not long-lived. By the time they are fifteen years old, the trees begin to die. As the old trees fall down, the musangas gradually disappear and saplings of other species gradually fill the gaps in the canopy. Within twenty years, the musanga community is replaced by a mixed secondary forest of many different species.

But this forest is still quite unlike the original jungle. The trees are much shorter—seldom more than ninety feet high—and all are relatively fast-growing species. Unlike the slow-growing, heavy-seeded trees of undisturbed jungle, most of them have seeds that are distributed by birds, bats, or the wind. Although vines are still plentiful, the undergrowth is thinner than in the earlier stages of revegetation, and it is no longer as difficult to move about as it was in the rampant growth of the first few years.

### *Biological nomads*

Outside Africa, jungle clearings are recolonized in much the same way, though by different kinds of trees. In South America the most important species in the early stages are various kinds of cecropia trees that look very much like the musanga and are in fact closely related to it. Soon after corn or cassava fields are abandoned, the first growth of bushes and vines is quickly replaced by huge numbers of cecropias, which sometimes grow in pure stands that look as if they had been planted. Like the musangas, the short-lived cecropias dominate the forest for only one generation. With no young trees to replace them as they die, the ground is soon occupied by a forest of many different species.

**Against a backdrop of tall jungle trees, vigorous cecropias spring up at the edge of an abandoned South American clearing. In a year or two, the entire opening probably will be covered by these vigorous pioneering trees of the second-growth forest.**

In Panama, Costa Rica, and other parts of Central America, on the other hand, one of the commonest colonizers of clearings and abandoned fields is the balsa tree, well known for its extremely light wood. And in the Philippines and Malaya, the major colonizers often are several small fast-growing species of *Macaranga.*

All these trees, whether related or not, have certain features in common. All of them grow rapidly—much faster than the giant trees of the true jungle—and all are soft-wooded and short-lived. They have efficient means for long-distance dispersal of their seeds, usually by birds, bats, or the wind. And since the seeds can remain dormant in the soil for a long time, they quickly take over when a clearing is made or a field abandoned. But since they cannot survive in deep shade, they cannot grow in the tall jungle unless some sort of gap is created.

Such trees are often known as "biological nomads"; under natural conditions, before man began to make clearings, they must have led a wandering existence, springing up on landslides, on sandbanks in rivers, and in gaps where old trees had fallen down. In the end, they are always suppressed by the slower-growing, more shade-tolerant trees of the tall jungle. But all of them, over the course of time, have acquired similar characteristics fitting them very successfully for life in jungle clearings.

## *New forests, new animals*

As these new forests take over in clearings, living conditions gradually become more junglelike. By the time they are forty years old, many of the trees may be well over one hundred feet tall. Once again the ground is shady, and the soil surface becomes cool, moist, and covered with litter.

But, because there are not nearly so many kinds of plants and animals, there is still a much narrower range of living conditions and a much smaller variety of niches to be occupied. The invading trees, for instance, usually have smooth

**Securely anchored by the stout claws on its hind feet, a three-toed sloth munches a giant leaf of a cecropia tree. As abandoned fields are colonized by second-growth forests, many other animals move in to take advantage of the new kinds of food that become available.**

bark; even when they are old and very large, they offer less of a foothold for epiphytes than did the original jungle trees. With fewer epiphytes, many of the ants and other small animals usually associated with air plants have difficulty finding places to live.

Even so, certain types of food are more plentiful than in the mature rain forest. Trees such as the cecropias and musangas produce great quantities of fruits that are very attractive to birds and fruit bats, and other trees have berries that are eagerly eaten by pigeons and many smaller birds.

But overall, animal life remains less varied than in the mature rain forest. Many of the new kinds of animals that move into the young second-growth jungle are quite different from those that inhabit undisturbed jungle. Three-toed sloths, for example, actually prefer these young forests with their abundant supply of cecropia leaves.

When the coarse grass, *Imperata*, invades jungle clearings, it is almost impossible for trees to become reestablished on the land. The grass's stems and tough root system quickly form such dense, tangled mats that seedlings of trees and other plants cannot survive.

## *The end of the cycle*

We know relatively little of what happens next in the mixed forests that eventually take over in clearings. In the face of exploding human populations and the constant need for more farmland, the young forests are rarely left undisturbed. By the time they are fifteen or twenty years old, the soil usually has recovered sufficiently to support a second food crop. The trees are felled and the cycle begins again.

In heavily farmed areas, even while the trees are growing up, the young forests are frequently damaged by grazing, by cutting timber, and occasionally by fire. Natural development of the forest often becomes so seriously disturbed that plants of open areas such as the bracken fern and the coarse grass *Imperata* invade the clearings and make it almost impossible for forest trees to become established. Large areas of wasteland that can no longer be used for farming of any sort are being created in this way in many parts of the tropics.

Yet if the replacement forests were left alone for a long enough time, they probably would eventually develop into tall jungle like the forest that was originally cleared. The mixture of trees that replaces the musanga community in Africa, for instance, includes some species that were found in the original jungle. If left alone for perhaps 150 to 200 years, the forest most likely would gradually become indistinguish-

In the early stages of regrowth on old clearings, young forests like this one in Brazil often form tangles that are even more difficult to clear than was the original jungle. If undisturbed, such forests eventually develop into something resembling the original jungle. But frequently the second-growth forests are cut down after a few years and crops are planted once again on the land.

Just as abandoned jungle clearings eventually are recolonized by jungle trees, disturbed land in other areas is colonized by a succession of plant communities leading to the stable climax community for the area. In New England, abandoned fields are taken over first by weeds and shrubs and then by white pines, which eventually are replaced by mature forests of maples, beeches, and birches . . .

able from virgin jungle. Only a few traces of farming people, such as fragments of pottery and charcoal in the soil, would remain as evidence that the land had ever been cultivated.

We know this is so, for much the same thing happens to disturbed land in other areas of the world. On abandoned farmland in New England, for instance, the ground is covered at first by thickets of bracken fern, sumac, and other bushes. These are later followed by stands of white pine which, if left alone, are replaced in time by forests of mixed hardwoods such as maples, beeches, and birch trees—probably very similar to the forests that covered these lands when the first colonists arrived. As in the jungle, the first, quick-growing communities of light-demanding species are gradually replaced by mixed communities of slower-growing trees whose seedlings are more shade-tolerant.

This process of gradual replacement of one community of plants and animals by another over a period of time is, in fact, a worldwide phenomenon, and ecologists give it a special name—*succession.* The process comes to a halt only when the land is occupied by what is called a *climax community,* the relatively stable, self-sustaining association of the living things that is best suited for survival under existing conditions of soil, climate, and so on. In many parts of New England, the climax community is mixed hardwood forest; in hot, humid, tropical lowlands the natural climax to succession is mature tropical rain forest.

## *Artificial ecosystems*

In addition to the shifting cultivation we have been discussing, large tracts of land that once were covered by rain forest now are occupied by permanent agriculture. In Malaya, mile after mile of landscape is dominated by the monotonous green of rubber trees. In Ghana and Nigeria, cacao farms and plantations of oil palms are nearly as common. And all over the tropics, acres and acres of jungle have been replaced by huge plantations of bananas, coffee, citrus fruits, tobacco, and other crops. Centuries ago, even the long-established, irrigated rice fields of southeast Asia were largely swamps and jungles.

Most of these plantation crops originally were members of the jungle community. The rubber tree still grows wild on riverbanks and in the periodically flooded forests along the Amazon and its tributaries. And cocoa, though it was already being cultivated in Central America when the first Spanish explorers arrived, is believed to have come originally from rain forests on the eastern foothills of the Andes, where cacao trees still grow wild in the understory.

But on plantations, these trees are grown under a completely new set of living conditions. Instead of living as scattered individuals in mixed communities of many kinds of trees, they usually are grown in almost pure stands. The plantations form what amount to artificial, man-made ecosystems, which are very different from the native jungle.

In rubber plantations, no other trees are allowed to compete with the crop. The only other vegetation permitted to grow are cover crops of grasses or vines which flourish in the light shade beneath the rubber trees. Besides preventing soil erosion, some of these plants—especially vines of the pea family—help maintain soil fertility: like clover, their roots have nodules containing nitrogen-fixing bacteria.

With just one kind of tree and a layer of low vegetation at ground level, the rubber plantation obviously has a much simpler structure than the jungle. And it provides a much smaller variety of ecological niches. Even in poorly tended, unweeded plantations, the total number of species of plants and animals is much less than in the jungle.

This artificial ecosystem also is less well balanced than a natural one. With fewer enemies to control their numbers, some organisms may even become pests. Termites, cockchafer beetles, and larger animals such as squirrels, rats, and wild pigs frequently damage the trees.

Far more serious are the fungal parasites that sometimes cause disastrous diseases. One of the worst is the disease caused by the leaf-blight fungus. In tropical America, the original home of the rubber tree, this disease does so much damage that development of large-scale plantations has been almost impossible. Although the leaf blight has not yet spread to the Old World, plantations in Africa and Asia are frequently afflicted by other less serious diseases.

The plantation also differs from a natural ecosystem in that its mineral cycles can be maintained only by the constant addition of fertilizers. Rubber is made from milky latex that is tapped from the living trees, and even an average plantation may produce several hundred pounds of dry

. . . In the Wasatch Mountains of Utah, on the other hand, the weeds and shrubs that move in after forest fires are replaced by deciduous forests of aspen trees. In time, however, the seedlings of Douglas firs and other conifers become established and eventually replace the aspens. This coniferous forest will perpetuate itself as long as the environment remains unchanged.

WESTERN UNITED STATES

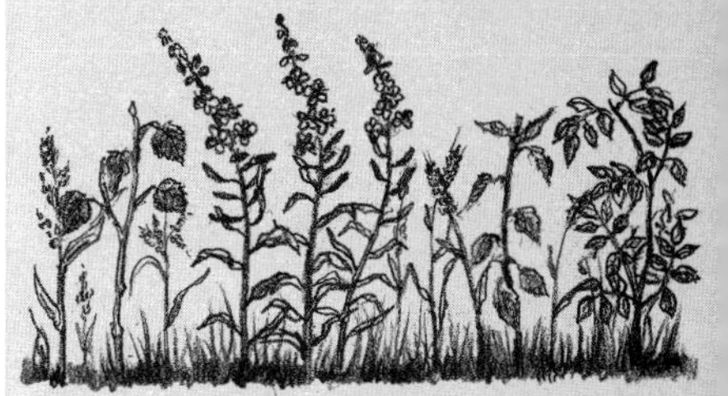

WEEDS AND SHRUBS

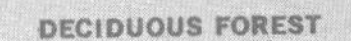

DECIDUOUS FOREST

CONIFEROUS FOREST

rubber per acre per year. But in addition to rubber, the latex contains rather larger amounts of various mineral elements. Nutrients also are lost in the drainage of rain water. Shortages even of elements the trees require in only minute quantities may be indicated by such things as discoloration of the leaves. Thus, unless nitrogen, phosphorus, potassium, and other elements are regularly added to the soil, the yield of latex soon begins to fall off.

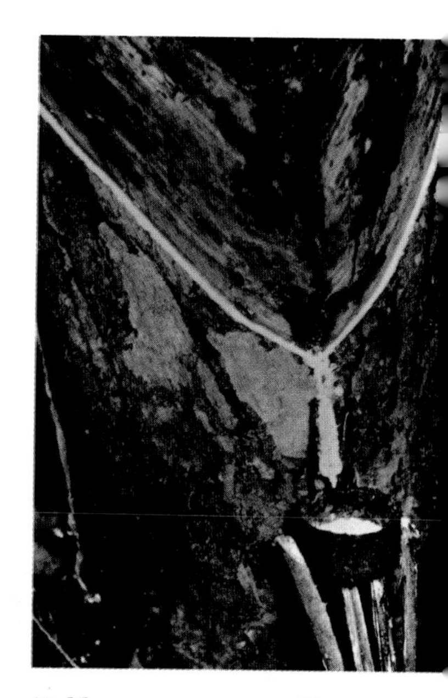

Rubber trees are tapped by cutting shallow grooves in the bark on the trunk. Within an hour, about an ounce of precious latex oozes from the wound and drips into a collecting cup. Each tree is tapped every other day and may yield twenty pounds or more of latex over the course of a year.

### *Cacao plantations*

A cacao plantation also is an artificial ecosystem, though at first glance it looks more like a natural forest than a rubber plantation does. The cacao trees grow only about twenty-five feet high and branch near the ground. Although they do not require shade in cultivation, they usually are planted beneath a scattering of much taller trees which provide shelter from the wind, and litter to the soil, and possibly benefit the cacao trees in other ways. On African cacao farms, the shade trees usually are survivors from the original forest that were deliberately left unfelled. But in Trinidad and other Caribbean islands, the land is normally cleared completely and special shade trees planted along with the crop. The tree most commonly used for this purpose is the beautiful scarlet-flowered immortelle, a kind of erythrina.

No ground cover is planted beneath the cacao trees. They are so close together and cast so much shade that little will grow beneath them in any case. And the few weeds that do spring up usually are left alone. But with three stories of vegetation—shade trees, cacao trees, and sparse undergrowth—the cacao plantation probably includes more niches for associated plants and animals than a rubber plantation does.

Yet it is far from being a stable or balanced ecosystem. As on rubber plantations, cacao crops are constantly plagued by pests and diseases, such as various viruses and the dreaded black-pod, due to a fungus that causes the fruit pods to turn black and rot. The plantations also are damaged by animals that are comparatively harmless in well-balanced

**The carefully tended rows of trees on a rubber plantation in Brazil look far different from the rain forest that once covered the land. Throughout the tropics, great tracts of jungle have been transformed into rubber plantations.**

## FOOD FOR THE GODS

The cacao tree, source of commercial chocolate, originally grew wild in Central and South American rain forests. When Europeans discovered America, the Mayans and Aztecs of Mexico already were cultivating the tree, which they considered food for the gods. Columbus brought a few pods back to Europe as a curiosity and chocolate soon became a valued luxury, first in Spain and later in Italy and other countries. As a result, cacao plantations sprang up in many parts of the tropics.

The chocolate itself is produced from the lima-bean-sized seeds contained in the large thick-skinned fruits. The seeds, called beans, are extracted from the pasty flesh inside the fruits, fermented to remove any remaining pulp, then washed, dried, and polished. At the chocolate factory, the beans are roasted, shelled, and then ground between heated rollers to extract melted raw chocolate, which is further refined and sweetened before use.

At the left, the trunk of a cacao tree is decorated with several fruits. On the right, workers on an African cacao plantation harvest their valuable crop.

communities. In the jungle, for example, the tropical American leaf-cutting ants spread out their activities so well that they seldom cause irreparable damage to individual trees or to any particular species. In cacao and citrus plantations in the West Indies and elsewhere, on the other hand, the ants often become major pests. There probably are several reasons for this. But the extent of their damage in plantations no doubt results in part because the ants have a smaller choice of plants to feed on and because the total biomass of vegetation is much smaller than in undisturbed jungle.

All in all, the ecological balance in a plantation is a precarious one. It can be maintained only as long as weeding, fertilizing, and pest-control measures are continued. If the plantation is left to itself, the cultivated crops are soon smothered by dense tangles of weeds, vines, and young trees of other species. The rubber or cacao trees may survive for a few years, but eventually they give way to a whole succession of new plant communities, similar to those that take over in abandoned cassava patches or rice fields.

### *Is the jungle doomed?*

Now that bulldozers and power saws are replacing axes and machetes, the jungle can be destroyed more rapidly and efficiently than ever before. As ever-increasing human populations clear more and more land to grow food and export crops, the jungle continues to shrink steadily in area. It is true that in remote areas such as in parts of New Guinea, Borneo, and the Amazon basin, some quite large areas of undisturbed jungle still exist. But if present trends continue even they may vanish by the end of the century.

This does not mean that the tropics then will have no forests at all. Relics of the virgin jungle probably will survive on steep slopes and in other inaccessible places, and there will no doubt still be large areas of second-growth forest. Some land will also be kept as forest reserves, but they will be managed for timber production rather than to preserve the native plants and animals. Some of the jungle will also be saved in national parks and protected wilderness areas such as the Henry Pittier Park in Venezuela, the Albert Park in the Congo, and the Bako National Park in Sarawak. In addition, a few jungle areas have been set aside for scientific research, such as Barro Colorado Island in the Panama Canal

A skid pole, elaborately guyed and rigged with block and tackle, is used to skid heavy logs out of the rain forest in Ghana. As worldwide demand for high-quality tropical hardwoods continues to grow, power saws and other mechanical aids are replacing axmen and hauling gangs throughout the tropics, and rain forests in all areas are being stripped ever more rapidly of their oldest and most valuable trees.

Zone, Finca La Selva in Costa Rica, and the University of Malaya's forest near Kuala Lumpur.

Yet if these tiny remnants of true jungle are all that are to remain, how much of the jungle's wonderfully varied life is likely to survive? There may still be vast areas of second-growth forest. But many jungle plants and animals do not live in such forests. Although in time these forests could change into a climax more or less like virgin jungle, this can happen only if true jungle exists nearby to serve as a reservoir supplying animals and seeds to restock the cleared areas. In places like southeastern Nigeria and the rice-growing areas of southeast Asia, the forest has been cleared for so long and over such large areas that it is doubtful if anything like the original jungle could ever return.

Nor do we know how large a reserve must be if it is to survive. An ecosystem can maintain itself only if it occupies enough area, and for tropical rain forests, this may be a matter of tens or even hundreds of square miles. Chimpanzees, for example, cannot maintain stable populations unless they are able to wander over extensive foraging areas, and the large carnivores need even more space. Many jungle trees also are widely dispersed and need large areas to maintain themselves. In Malaya, one researcher found that, for some species, no more than one individual grows on every seven and one-half acres of land.

## *The future of the jungle*

Certainly large areas of the jungle are doomed to disappear. Human populations in many tropical countries are desperately short of food and even of living space. The jungle, with its wealth of plant and animal life, is bound to be sacrificed to meet these needs. Yet some areas of undisturbed jungle still can be saved. With improved agricultural methods, much of the land that has already been cleared certainly could be made more productive.

Even so, some people would still prefer to eliminate the jungle entirely, for it is a major reservoir for some of the

**In Asia, huge tracts like this rice plantation in Ceylon have been under cultivation for so long that it is doubtful if rain forest could ever become reestablished on the land.**

most dreaded diseases that attack crops, domestic animals, and even man himself. The yellow fever virus, for instance, maintains itself in the blood of monkeys and other jungle animals. Yet many of these disease organisms are also harbored in animals of second-growth forests. If we were to destroy all the rain forests in an attempt to eliminate these diseases, it is possible that they would still survive in marginal wastelands and second-growth forests.

Yet the jungle has many positive values. Like any forest, it protects the soil against erosion, especially on steep slopes. It is the source of some of the world's most beautiful and valuable hardwoods, such as mahogany, teak, greenheart, iroko, and meranti. In addition, scores of other tropical timbers probably are nearly as useful as these trees, but at present they are little appreciated on the world's markets. Many drugs, dyes, insecticides, and other valuable plant products also come from jungle trees and vines. And who can guess how many other plants may in time prove useful?

Besides its economic values, the jungle is worth saving for compelling scientific reasons. In the past, the tropics have contributed an immense amount to man's knowledge of nature. Much of what we know about the evolution of plants and animals and even of man himself was learned by naturalists working in tropical forests. A great deal of our knowledge of such things as mimicry, animal behavior, and the endlessly varied and sometimes astonishing relationships between plants and animals also has been gained in the jungle. And many secrets are yet to be unraveled in these complex natural laboratories. If the jungle is destroyed before we have had a chance to study it, whole chapters in biology may never be written.

But most of all, the jungle is a place of mystery and beauty. To visit the rain forest is to be overwhelmed by the variety and complexity of a unique living world of nature that has flourished for millions of years. Why save the jungle? Come with an observant eye and the question will answer itself.

**Like a symbol of the unending renewal of all life, a cecropia seedling grows on the branch of a fallen jungle tree, spreading its tender new leaves where sunlight penetrates a gap in the forest canopy.**

# *Appendix*

# *Jungle and Junglelike Preserves of the United States*

Since all the continental United States lies within temperate and arctic regions, true tropical rain forests on United States territory exist only in Puerto Rico and, to a lesser extent, in the Virgin Islands. But some very interesting subtropical cloud forests are found on certain mountain slopes in Hawaii, and even in the southernmost portions of Florida and Texas, which come close to the tropics; there are some truly junglelike areas where tropical plants and animals can be found.

In all these areas, significant jungle preserves have been set aside, both as national parks and wildlife refuges maintained by the United States Department of the Interior and, in Puerto Rico, as a national forest administered by the U.S. Forest Service of the Department of Agriculture. In these preserves, museum exhibits and lecture programs by trained naturalists orient the visitor and help him interpret the ecology of the rain forest. And miles of well-maintained trails provide unexcelled opportunities for first-hand observation of the intriguing plant and animal life and captivating scenery of the jungle.

BAMBOO

*Caribbean National Forest (Puerto Rico)*

This 12,500-acre tract, the only tropical area maintained by the U.S. Forest Service, is located in the Luquillo range of mountains in northeastern Puerto Rico. El Yunque, a 3486-foot-high peak, is the highest point in the forest. Readily accessible by modern highways, the area is visited each year by thousands of tourists who come to enjoy spectacular mountain scenery and the luxuriant vegetation of an extensive rain forest laced with rushing streams and crisscrossed by many miles of hiking trails. Tree ferns and bamboo are among the more popular plant attractions. The preserve is also the only home of a remnant population of the rare, endangered Puerto Rican parrot. (See pages 61 to 65.)

*Everglades National Park (Florida)*

This huge park of nearly two thousand square miles at the southern tip of Florida is the largest remaining subtropical wilderness in the United States. Most of the park is a vast watery marsh filled with sawgrass and fringed by mangrove swamps. But here and there the flat expanses of the park are dotted by low islands—none of them over ten feet high—called hammocks. The islands are overgrown with stately palms, mahoganies, strangling figs, and other West Indian trees and shrubs that flourish in junglelike profusion. In some places, elevated boardwalks permit visitors to wander through these areas and marvel at the wealth of epiphytes, especially orchids, as well as colorful birds, butterflies, and occasionally a few alligators and crocodiles.

*Hawaii Volcanoes National Park (Island of Hawaii, Hawaii)*
Beneath Kilauea, one of the most active volcanoes in the world, mountain slopes on the windward side of the island are covered with tropical rain forest near sea level. At higher elevations this wet jungle merges into a dry transition forest and then into a mountain cloud forest where, in some areas, the annual rainfall is between 100 and 180 inches per year. Here there are extensive forests of tree ferns, as well as ohia-lehuas, sandalwoods, and other unusual trees. Apapanes, amakihis, iiwis, and many other incredibly colorful native Hawaiian birds also find refuge in the park.

TREE FERN

*Santa Ana National Wildlife Refuge (Texas)*
Situated on the Texas-Mexico border in a bend of the Rio Grande, this wildlife refuge of approximately two thousand acres includes many junglelike features and a strange mixture of temperate and tropical trees. More than three hundred bird species have been observed in the refuge, including many typically Mexican species which are not found further north, such as rose-throated becards and kiskadee flycatchers. The varied mammal life includes animals such as armadillos and rarely observed ocelots and jaguarundis.

*Virgin Islands National Park (Virgin Islands)*
This tropical island park in the Caribbean east of Puerto Rico includes about two-thirds of St. John, the smallest of the three principal United States Virgin Islands. Prime attractions are the park's gleaming white sand beaches and offshore coral reefs. In the uninhabited interior, where volcanic mountains rise to 1277 feet, are impressive tropical forests. In colonial days, much of the island was planted in sugarcane, but now the plantations have been abandoned and haunting ruins of the estates dot a landscape that has been recolonized by tropical hardwood forests of gumbo limbos, mahoganies, and other tropical trees. Epiphytic bromeliads and orchids are a common sight along the many roads and trails that penetrate the interior jungle.

# *Jungle and Junglelike Preserves of the World*

Since the establishment in 1872 of Yellowstone National Park, the first such preserve in the world, the national-parks movement has spread to many countries. As a result, large tracts of many types of habitats—including jungles—have been set aside by various governments to preserve and protect unique scenic, plant, and animal resources.

It is important to remember, however, that in other countries the definitions and uses of national parks often differ substantially from the American concept of such a reserve. Africa has a number of parks of excellent quality and great size. India also has a few outstanding areas. In southeast Asia and Central and South America, on the other hand, national parks have been slow in developing and may not offer the same sort of facilities and controls that we have come to expect of national parks in the United States.

This is due in part to the fact that most tropical countries are poor and underdeveloped. And in many of them, efforts at conservation education and legislation have been sorely inadequate or even lacking entirely. Taking these things into account, the traveler should be grateful for any park areas that do exist. By showing his sincere appreciation for the (economic, esthetic, and scientific) values of existing parks and by convincing local people of the need to preserve more of these resources, any visitor to foreign parks can help to further the cause of international conservation.

Because of the vast areas and great number of countries involved, only a few of the wildlife preserves of the tropics can be mentioned here. The following list summarizes the attractions and facilities of some of the outstanding jungle sanctuaries around the world.

OKAPI

## AFRICA

### *Congo*

*Albert National Park,* situated between Lake Kivu and the Ruwenzori Range of mountains, is a spectacular scenic area of 3124 square miles. This topographically varied preserve includes two separate tracts of jungle. One area is centered on the basin of the Middle Semliki River and its tributaries, where until very recent times tribes of pygmies have lived in complete isolation.

Here too one may glimpse okapis, the striped, horse-sized relatives of giraffes that were not discovered until 1909. The park's other rain-forest area is on the base and flanks of the Ruwenzori Range itself and is inhabited by animals such as forest elephants, dwarf buffaloes, chimpanzees, and Ruwenzori colobus monkeys. This magnificent but relatively isolated park is difficult to enter and tourism has not yet been developed.

***Ghana***

*Mole Game Reserve* is an excellently managed, 1600-square-mile sanctuary in northern Ghana. Most of the park is rolling savanna, bright with flowers at the end of the wet season, but there are strips of thick jungle along streams. The park's wealth of animals includes lions, elephants, warthogs, aardvarks, baboons and several other kinds of monkeys, and many unusual birds. Except in the wet season, the park is easily accessible by road from the town of Tamale. A motel and small chalets provide good accommodations.

FOREST ELEPHANT

***Ivory Coast***

*Mt. Nimba Nature Reserve* is a virtually uninhabited wilderness area protected by the Ivory Coast Forest Service and the Institute of French Black Africa. The reserve, located in the Ivory Coast but also extending into Liberia and Guinea, totals over 25,000 acres and has dense cloud forests, chimpanzees, and abundant bird life. A small scientific research station is situated on Mt. Nimba in the Guinea sector of the reserve. A few jeep roads penetrate the area, but tourism is not developed, and a permit is required to enter the reserve.

***Kenya***

*Meru Game Reserve,* while not a national park, has many special attractions for the tourist. This was the home of Elsa, the lioness, as described in Joy Adamson's books and films. Rain forest exists only along the main rivers and in the Ngaia Forest. Although this is a vast wild country of 250,000 acres, there are a few roads where wildlife is easily observed and birds are colorful and abundant. Accommodations are available at one hotel and a few shelters and campsites. Visitors can reach the preserve by two roads from Nairobi or by plane to two local air strips.

***Tanzania***

*Gombe Stream Game Reserve* on the northeastern shore of Lake Tanganyika was established mainly for the protection of chimpanzees. In the reserve's sixty-one square miles of lake shore, rugged mountains, and dense forest, the visitor may also see red colobus monkey, bush bucks, leopards, sunbirds, and other notable wildlife. The reserve can be reached by boat from Kigoma.

*Kilimanjaro Game Reserve,* a sanctuary of 720 square miles in northeastern Tanzania, includes much of Mt. Kilimanjaro between six thousand feet and the peak's snowclad summit. The cloud forest at higher elevations harbors duikers, forest elephants, colobus and blue monkeys, and abundant bird life. There are plans to make the reserve a national park.

*Mt. Meru Game Reserve,* an area of ninety-nine square miles also on Mt. Kilimanjaro, includes rain forest for a short distance above 5500-feet elevation. Giant forest hogs, colobus monkeys, and many mountain birds attract visitors despite the fact that access is only by one roadway and hiking trails.

### *Uganda*

*Bawamba Forest,* an ecological unit of the great Ituri Forest of the Congo, includes extensive stands of ironwood trees and typical lowland jungle. While traveling along footpaths, elephant and buffalo trails, and a few roads, the visitor may see otter shrews, bush babies, galagos, colobus monkeys, African palm civets, golden cats, a few chimpanzees, and almost four hundred species of birds.

AFRICAN CUCKOO FALCON

*Budongo Forest,* in western Uganda between Masindi and Lake Albert, is one of the most accessible areas for viewing chimpanzees, as well as scaly-tailed flying squirrels, tree pottos, two species of duikers, giant forest squirrels, and black and white colobus monkeys. In addition to networks of forest roads and tracks, the area offers accommodations in a guest house.

*Kayonza Forest,* a nearly impenetrable tract of jungle carved by deep gorges, merges at higher elevation with upland rain forest. In addition to many primates, the forest has hammerhead fruit bats, pottos, African striped weasels, African civets, and many interesting birds, including turacos and cuckoo falcons.

*Mpanga Forest* is a vast tract of rain forest north of Lake George, partially opened up by timber roads. Notable wildlife includes elephants, colobus monkeys, chimpanzees, galagos, and scaly-tailed flying squirrels.

*Murchison Falls National Park,* 1505 square miles in extent, includes small areas of jungle growth and strips of tropical forest along the Victoria Nile River. Here too chimpanzees are a prime attraction.

*Queen Elizabeth National Park* is an area of over seven hundred square miles between Lake Edward and Lake George. The park boasts a remarkable range of habitats, including tropical rain forest in the Maramagambo Forest. Driving through this area, the visitor may see chimpanzees, blue monkeys, red-tailed monkeys, and a few rare red colobus monkeys. The popular chimpanzee lookout tower is one of the best places to observe these interesting apes. In addition to a few campsites, hotel accommodations for about seventy people are available in the park.

## ASIA

### *Ceylon*

*Gal Oya Valley National Park,* a ninety-eight square-mile reserve, is a mixture of savanna and dry evergreen tropical forest. Birds are abundant, as are leopards, pythons, cobras, deer, and elephants. After flying in from the capital city of Columbo, access is by foot or riverboat; tourist accommodations are limited to two small rest houses.

*Ruhuna National Park* is primarily thorny scrubland, but tropical forests border some of the rivers and lakes. In the park's ninety square miles, over two hundred species of birds have been sighted, along with chital deer, elephants, and sloth bears. The park is traversed by trails and thirty miles of good gravel roads. A few bungalows provide lodging for visitors.

### *India*

*Corbett National Park,* near New Delhi, is India's oldest national park. The area includes 125 square miles, ranging from the Ranganga River plains to elevations of over four thousand feet in the forested foothills of the Himalayas. Tigers and great Indian hornbills are the park's most notable attractions. Comfortable tourist facilities are available, including elephants for riding.

*Hazaribagh National Park* is a densely forested, seventy-one square-mile refuge where tigers, bears, wild pigs, and many jungle birds abound. No tourist facilities are available.

*Kaziranga National Park* is a heavily vegetated 166-square-mile area stretching for twenty-five miles along the Brahmaputra River. Strict controls are maintained to preserve the park's wildlife, which includes a small but growing colony of the rare Indian rhinoceros, as well as wild pigs, tigers, leopards, and bears.

### *Indonesia*

*Udjung Kulon Nature Reserve,* on Java, is a tract of over 100,000 acres offering a tremendous variety of plants and animals. The area is famous as the last stronghold of the little Javan one-horned rhinoceros; with only thirty individuals remaining, this is considered to be the world's rarest large mammal. In addition,

JAVAN RHINOCEROS

the park is a refuge for tigers, barking deer, and teladus ("stinking badgers"). Tourist facilities are very limited.

***Malaysia***

*Bako National Park,* in the state of Sarawak, is the most accessible nature preserve on the island of Borneo. It is located on the coast only a few miles from the capital city of Kuching and is easily reached by boat. Although it includes only ten square miles, the park is noted for its scenery, with a mixture of tropical forest and open shrubland, and magnificent views of the ocean and the fantastic peak, Santubong. The chief botanical attractions are pitcher plants, rhododendrons, orchids, and strange epiphytes such as *Dischidia* and *Hydnophytum.* The most notable animals are gibbons, but there are also wild pigs, mouse deer, several kinds of monkeys, including the shy and odd-looking proboscis monkeys, and a great variety of birds. The park has a good system of trails, and a rest house and two bungalows are available for visitors.

*Mount Kinabalu National Park,* in the state of Sabah on the island of Borneo, includes the highest mountain in southeast Asia (13,455 feet). With an estimated population of 2500 orangutans, the park is one of the last strongholds of these endangered apes. The extremely rare Sumatran two-horned rhinoceros still lives in parts of the park, and there are gibbons, barking deer, tree shrews, and other unusual mammals. The most remarkable plant is *Rafflesia,* a parasite noted for its huge blossoms. The cloud forests abound with rhododendrons, tree ferns, orchids, and more and larger pitcher plants than are found anywhere else in the world. The park is accessible by road from the coastal town of Jesselton, but higher elevations can be reached only on foot. Accommodations include a rustic hotel and four mountain cabins.

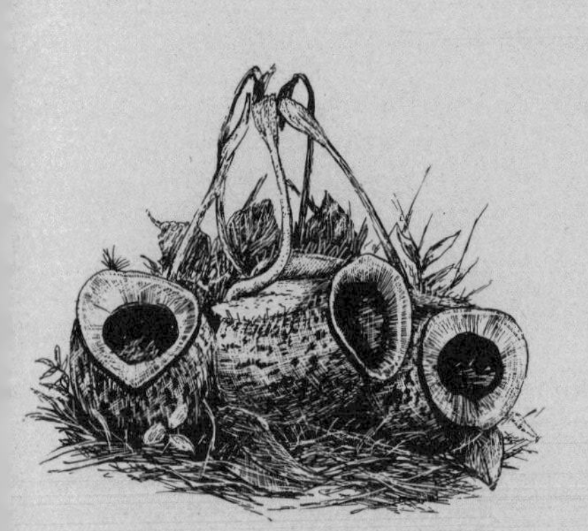

PITCHER PLANT

*Taman Negara National Park* (formerly known as King George V National Park) includes 1760 square miles of beautiful mountain country on the Malay Peninsula. The park is covered with dense tropical rain forest, inhabited by tigers, Malayan tapirs, slow lorises, and many kinds of monkeys. Visitors can enter the park on trails or by boat and find limited accommodations in river lodges.

***Philippines***

*Mount Apo National Park,* on the island of Mindanao, is a conservation area that covers 297 square miles and reaches altitudes of 9700 feet. In the diverse habitats on the mountain slopes, notable animals include flying lemurs, several species of bats, flying squirrels, palm civets, and the very rare and endangered Philippine monkey-eating eagle.

### *Thailand*

*Khao Yai National Park,* a heavily visited, eight-hundred square-mile reserve not far from Bangkok, contains superb tropical forests, valuable water resources, and varied wildlife. Although some jungle species such as the Javan and Sumatran rhinoceros and the rhesus monkey have either disappeared from the park or become very rare, strong efforts are being made to preserve the remaining animals such as tigers, leopards, bears, langurs, and sambar deer. Access is by car, plane, or on foot, and the park provides camping areas and bungalows.

RHESUS MONKEY

## AUSTRALIA

*Bellenden-Ker National Park,* in southern Queensland, has many jungle-clad peaks over four thousand feet in elevation. Orchids and vines are plentiful, and two hundred species of birds have been recorded. In addition to traces of ancient aboriginal tribes, the visitor can see several kinds of marsupials including kangaroos. Tourist facilities are lacking.

*Bunya Mountains National Park* in the Great Dividing Range of mountains has both wet and dry tropical vegetation. Especially notable are the *Araucaria* forest and the stands of wild yellow stringybark trees. Kangaroos are abundant. Tourist facilities are limited.

*Daintree Gorge National Park,* also located in the Great Dividing Range at elevations of about five thousand feet, contains jungles, deep gorges, and beautiful waterfalls, such as the famous Adeline Falls. There are no tourist accommodations.

*Eungella National Park* is Queensland's largest park, with over 120,000 acres of tropical forests, mountains, gorges, and streams. Palms and vines grow luxuriantly. The area also has many birds, of which one of the most unusual is the grey swiftlet. Tourist facilities are planned.

## CENTRAL AMERICA

### *Guatemala*

*Tikal National Park* is primarily an archeological zone featuring partially restored ruins of a great religious center of the pre-Colombian Mayan Indians. The surrounding lowland rain forest offers many attractions for visitors interested in natural history. From the tops of impressive temples that loom up through the trees there are superlative vistas across the vast canopy of the

Petén jungle. Along the forest trails are creatures such as pumas, jaguars, agoutis, deer, tapirs, ocelots, and hundreds of species of birds. The park is accessible by plane or on horseback, and rustic lodgings are available.

***Panama***
*Barro Colorado Island* is primarily an area for the scientific study of Central America's rapidly vanishing wildlife. It is not a national park, but is operated as a biological research station by the Smithsonian Institution. The preserve is located on an island that was formed by the rising water of Gatun Lake during construction of the Panama Canal. The island, about three miles in diameter and accessible only by boat, is covered with seasonal dry forest crisscrossed by numerous trails. In the forest are many colorful tropical birds, and seventy-four species of mammals, including howler monkeys, two-toed and three-toed sloths, jaguarundis, pumas, ocelots, peccaries, and brocket deer. Modest living accommodations are available but usually are occupied by scientists. Permission to visit the reserve must be secured in advance from the director of the Research Station.

OCELOT

## SOUTH AMERICA

***Brazil***
*Itatiaia National Park*, somewhat over 30,000 acres in extent, was created around an old biological research station affiliated with the fine Botanical Gardens in Rio de Janeiro and is easily reached by road from that city or from São Paulo. In this magnificent park with a maximum elevation of about 9200 feet, the lower slopes are covered with a luxuriant forest filled with groves of tree ferns, lianas, orchids, begonias, and a profusion of brilliantly colored flowers. At about five to six thousand feet the forest merges with a cooler, shorter cloud forest, and the highest elevations are covered with a tussock grassland with many strange plants. Lizards, snakes, colorful frogs, jaguars, coatimundis, deer, and vultures are some of the wildlife attractions.

*Organ Mountains National Park*, located about one hundred miles northeast of Rio de Janeiro, includes over 25,000 acres of luxuriant jungle teeming with wildlife. Like Itatiaia, the park boasts exceptional mountain scenery, with fantastically shaped peaks rising to over 7200 feet. Especially notable is the famous pinnacle called God's Finger. Tourist accommodations near both parks are good to excellent.

*Tijuca National Forest*, situated almost within the city limits of Rio de Janeiro, offers a glimpse of a lush Brazilian forest even to tourists with very little time at their disposal. Located on the slopes of steep, jagged mountains behind the city, this eight-thousand-acre reserve was mostly covered by coffee plantations not many years ago. Now the land is cloaked with a luxuriant forest rich in lianas and epiphytes. From well laid-out roads and trails, the visitor can easily enjoy views of rushing streams, palms, begonias, bromeliads, and many orchids.

*Guyana*

*Kaieteur National Park*, located on a plateau slightly over one hundred miles inland from Guyana's Caribbean coast, is an area of forest and savanna along the banks of the Potaro, Mure Mure, and Elinku Rivers. These rivers merge and form the park's most famous feature, Kaieteur Falls, which provide a spectacular display as they drop vertically for 741 feet. Mammals are abundant in the forest, as well as characteristic jungle birds such as tinamous, toucans, and parakeets. The park is accessible by air or by boat on the Potaro River from the coastal city of Georgetown. Lodging for about thirty persons is available.

*Venezuela*

*Avila National Park* and *Guatopo National Park,* both located near Caracas, and comprising over 350,000 acres, were established in part to safeguard the water supplies of Venezuela's bustling capital city. Although they both include private agricultural holdings, they also contain some primeval rain forest with jaguars, deer, tapirs, pacas, peccaries, and rich bird life, including many parrots. Both parks are developed for tourism, and there is a biological research station at Guatopo.

*Henry Pittier National Park,* formerly called El Rancho Grande, is located just a short distance from Caracas and is easily reached by car. It contains almost 250 square miles, ranging from sea level to heights of about eight thousand feet and including both lowland rain forest and mountain cloud forest. Wildlife flourishes in this rich variety of habitats, including 530 species of birds and mammals such as pumas, jaguars, and capybaras. The park is the site of the famous biological research station, Rancho Grande, once frequented by the American naturalist William Beebe and still permanently staffed by a biologist. This readily accessible and heavily visited park includes hiking trails, fishing and picnic areas, campsites, lodges, and museum exhibits.

COLLARED PECCARY

# *Jungle Life in Danger*

Jungles throughout the world are continually shrinking in size. Every year hundreds of square miles of tropical forest are replaced by agricultural land, roads, airfields, and dwelling sites, and millions of trees are cut for lumber. But when the forest is felled, the whole ecosystem is destroyed. Soil fertility is depleted, surface soil is carried away by erosion, and living conditions are altered so radically that many kinds of plants and animals are unable to survive.

Some jungle creatures already have disappeared and, each year, other species are being nudged ever closer to extinction. The orang-utan, mountain gorilla, Sumatran rhinoceros, giant river otter, and many other large mammals have become exceedingly rare, as have some of New Guinea's beautiful birds of paradise. Even some of the jungle's trees are passing into oblivion. *Amherstia nobilis,* a large tree often cultivated in tropical gardens, has practically disappeared from its native habitat in the jungles of the Lower Salween Valley of Burma. In the West Indies, the mastic tree, *Mastichodendron sloaneanum,* once a source of valu-

**SUMATRAN TIGER**
**Uncontrolled hunting and trapping have seriously reduced the numbers of these magnificent animals, which now live only in the mountains of northern and southwestern Sumatra. Also endangered are other races of these cats—the larger Bengal tiger of India, the Javan tiger, and the smaller, brighter Bali tiger. Only complete prohibition of hunting and protection within national parks may save these big bold cats. A great deal of education also is needed to persuade the natives to conserve an animal that can be a dangerous man-eater.**

**PYGMY HIPPOPOTAMUS**
**This small-scale cousin of the hippopotamus is about the size of a large pig and weighs up to five hundred pounds. It is found only in rivers and swampy jungles of Africa's Ivory Coast and, in smaller numbers, in Liberia, Sierra Leone, and Nigeria. Hunting by natives, who prize its flesh, and trapping by zoos for exhibition purposes are partly responsible for its scarcity, but destruction of its habitat also contributes to the pygmy hippopotamus' continuing decline in numbers.**

able timber, has all but vanished. Two hundred years ago it was common in Barbados, but now it is extinct there, although a single specimen may still survive on the island of Antigua. In the Seychelles Islands, many unique jungle plants are now rare or extinct. The tree *Vateria seychellarum,* for instance, is the only member of an important family of jungle trees, the dipterocarps, found in the western Indian Ocean. Yet only three of these trees remain alive on the Seychelles.

Fortunately, however, many endangered jungle plants and animals can still be saved—if we act quickly. First of all, local people must be educated to the need for preserving their unique natural treasures. Hunting and trapping of tropical species must be regulated, and laws must be passed prohibiting the import and export of endangered wildlife. And there must be enough game wardens to strictly enforce protective laws. But most important of all, more and larger jungle preserves and wildlife refuges must be established to provide living space for *all* jungle wildlife.

Pictured below are several representative jungle plants and animals whose survival depends on prompt enactment of conservation measures.

CENTRAL AMERICAN TAPIR

**This curious, donkey-sized tapir lives only in the wildest rain-forest areas of Central America and northwestern South America. Men have always hunted the Central American tapir for its flesh, as do some of the larger predators, especially jaguars. But the clearing of jungles and draining of marshy areas are just as important in contributing to the animal's decline. Unless sanctuaries are established in the few remaining tracts of Central American jungle, this shy creature will surely be doomed.**

PYGMY CHIMPANZEE

**This small, glossy black ape, about half the size of a chimpanzee and more slender in form, lives only in the tropical rain forest on the left bank of the Congo River in Africa. Little is known of the life history of this quiet, gentle little chimp; indeed, it was not described scientifically until 1929. Collection of this rare species for zoos and exploitation of the Congo jungles for lumbering and cropland pose the gravest threats to the pygmy chimpanzee's survival.**

LAELIA ORCHID

The delicate, rosy-lavender orchid, *Laelia purpurata*, grows on rocky, sunny cliffs and mountains in the Brazilian coastal jungles near Rio de Janeiro and Santos. Since it produces many natural varieties, the species has long been a favorite with orchid fanciers as a subject for hybridization. But, like many other orchids in South and Central America, it has been virtually eliminated from its limited range by ruthless collectors. Only rigid restrictions on orchid hunters, exporters, and florists can save this and many other orchids from extinction.

IMPERIAL PARROT

This handsome parrot lives only on the island of Dominica in the West Indies. In the 1920s the bird was scarce, and by 1950 it was described as "rare." But now its very survival is at stake, for lumbering interests are about to cut all that is left of Dominica's unique rain forest, one of the few remaining virgin forests in the Caribbean. In addition, natives occasionally shoot and trap the bird. Strict enforcement of protective laws, scientific research, and establishment of a large forest preserve are desperately needed to save the imperial parrot.

QUETZAL

This beautiful emerald and ruby bird, the national emblem of Guatemala, boasts an extraordinary historical, religious, and anthropological fame in Mexico and Central America. But shifting cultivation by Indians and the establishment of large plantations are rapidly destroying the habitat of this bird, which cannot tolerate human activity and can live only in undisturbed forest. Unless preserves for it are set aside immediately and efficiently guarded by trained wardens, this magnificent bird may soon disappear.

# *Exploring the Jungle*

Camping and traveling in the jungle are basically not very different from roughing it in any other wilderness area. Although you must guard against a few hazards, the dangers are not nearly so great as most people imagine. By being sensible and observing a few simple rules, almost any outdoorsman can safely enjoy a few days or even weeks of camping out in the tropical rain forest. This is, in fact, the best way to experience the true grandeur of the jungle and to appreciate its fantastic profusion of life.

A basic rule for jungle travelers is never to wander about alone. Unless you are within sight and sound of your camp or are following a well-marked trail, always have at least one companion.

The best companion for a jungle trip is a native of the area you are visiting. Even when no path is visible, he will prove a reliable guide. In addition, he will know such things as how to build a comfortable, rainproof shelter from forest materials and how to make a fire even in pouring rain. In many cases, he will also be a mine of information about jungle animals and plants. Seek your guide at a farm or village in an isolated clearing, however; a man from a town or distant village probably will know no more about the jungle than you do yourself.

## *What to wear and what to carry*

In the hot, humid jungle climate, you need clothing that is light in weight, yet sturdy enough to protect you against insect bites and thorny plants. Wear long-sleeved shirts of tough cotton, denim, or chambray, and keep them tucked into your trousers. Like your shirt, your trousers should be of a strong, lightweight material not easily torn or pierced by thorns. They should have loose, wide legs and the bottoms should be tucked into your boots or socks to keep out ticks and crawling animals.

After dark you should change into clean, dry clothing that covers your arms and legs, since the danger from malaria-carrying mosquitoes is greater at night than during the day. Shorts, though comfortable, are not recommended for jungle wear. In some parts of the eastern tropics, it is actually dangerous to wear them because the mites that carry the dangerous disease, scrub typhus, readily crawl onto bare knees.

As for footwear, be sure to have several pairs of absorbent cotton or woolen socks. Avoid all-rubber boots since they make your feet hot and encourage fungal infections. Instead, experts recommend army jungle boots which come up to or over the calves. These boots, made of leather and canvas and fitted with cleated rubber soles, are much better than all-leather boots since they dry more easily and are less likely to mildew. Never walk barefoot, even in camp, and be especially careful to keep your feet covered near streams or rivers, where you could become infected with hookworm.

Although a hat is not necessary and may even be hot, many people find it comfortable to wear one that is lightweight and broad-brimmed. If it is waterproof, it will also provide some protection against rain. All other raingear should be of the lightest possible material, such as plastic or nylon; anything heavier becomes uncomfortably hot.

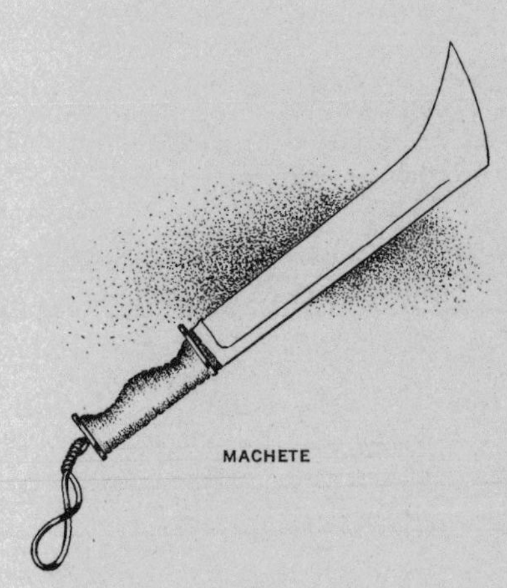
MACHETE

All clothing that has been worn should be washed each day and dried by the campfire. Store your clothing in strong plastic bags or other waterproof containers, but remember to open them regularly to air your clothes when the weather is warm and dry. Optical equipment such as binoculars and spare camera lenses also should be stored in airtight containers, with silica-gel desiccators to prevent damage by fungi.

To complete your outfit, you need a machete, the all-purpose tool of the tropics. Carry your camera, notebook, and small accessories in a waterproof haversack. If you have a heavy load and expect to walk long distances, your other gear is best carried in a lightweight, aluminum-framed backpack with a waterproof cover.

## *Making camp*

Because of the constant high humidity of the jungle interior, you should always set up camp in a clearing unless it is to be just one night's shelter. Never camp directly under large trees, especially if they have dead limbs, and avoid camping near insect-infested lakes and marshes.

Tents are undesirable for jungle use, since they become intolerably hot and stuffy. A tarpaulin that can be quickly spread over a frame of light jungle wood makes a much cooler shelter. However, in most jungles you need not carry even a tarpaulin; native guides generally can construct a waterproof shelter from small tree stems, palm leaves, and lianas, and they can build it almost as quickly as you could pitch a tent.

Do not sleep on the ground. Use a campbed or a jungle hammock, and enclose it well with a mosquito net. Besides protecting you from mosquitoes, the net will keep out intruders such as moths, flying beetles, and vampire bats. And remember that, even in the tropics, a warm blanket is usually welcome at night.

A campfire is always a necessity for boiling water, cooking food, and drying your clothing. It also helps discourage mosquitoes and curious animals, and makes for a cozy, safe atmosphere.

## *Food and drink*

You need not carry drinking water in tropical rain forests, since clear, fast-running streams usually are common. However, *all water must be boiled before use.* Near villages and farms, it is liable to be infected with bacteria and viruses, and even in remote areas may be contaminated with flukes and other parasitic worms. If, in an emergency, no boiled water can be obtained, the water should at least be treated with purification tablets. Your guide may also be able to find a species of liana that will provide safe, palatable water when the stems are cut.

As for food, your best choices are the readily available native staples such as rice, corn, yams, manioc, and beans. You can also buy pineapples, papayas, bananas, and other fruit from jungle farmers. If you hire a native hunter, he will prove very skillful at catching fish and shooting birds, wild pigs, agoutis, pacas, and other wild game. For variety, you might also carry some dried, vacuum-packed imported foods, but they are certain to be more expensive than local products.

Do not expect to find wild food on your own. Edible plants are extremely scarce in the jungle. Although some fruits and nuts are good to eat, many others taste unpleasant or are even poisonous. Do not experiment with unknown foods unless you first consult a reliable field guide. Better still, depend on the advice of your native guide.

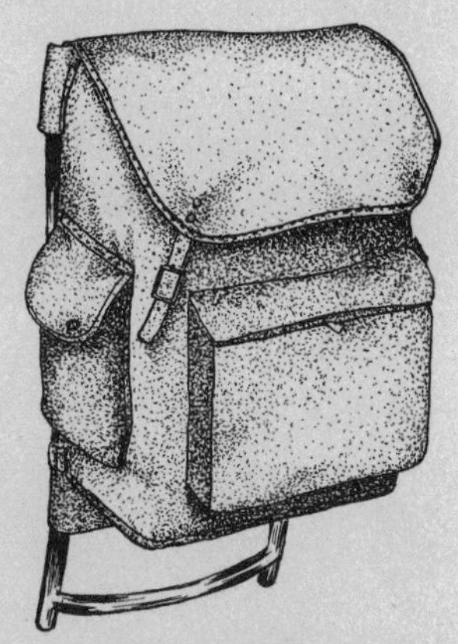
BACKPACK

## *Snakes and other dangers*

Although most jungle animals are harmless to man, a few of them can be dangerous. Snakes in particular should always be treated with respect. The best protection against accidentally stepping on a poisonous species is to wear high, stout boots at all times. And be prepared for the worst: carry snakebite serum, with instructions from a competent medical authority on how to use it.

Scorpions also can give painful bites, but they are seldom seen in the jungle. Tarantulas, which are more common, look terrifying but are in fact quite harmless to man. Ticks also are abundant in most jungles and are very difficult to discourage; they must be carefully removed at the end of each day. In Asian jungles, land leeches also are troublesome pests. Although they readily attach themselves to bare arms and legs, a lighted cigarette or a pinch of salt is usually enough to make them release their hold. The wounds where they puncture the skin should be carefully treated with antiseptics or antibiotics, as should all other bites and scratches: in the hot humid jungle climate, wounds are slow to heal and easily become infected.

As for large carnivores, such as jaguars and leopards, most are scarce and shy and very rarely attack humans. The greatest dangers, in fact, are not large animals but the small ones that carry diseases. Mosquitoes, for example, carry malaria, yellow fever, filariasis, and other diseases. The risk of contracting malaria is not great, provided you regularly take antimalarial drugs, always sleep under a mosquito net, and protect your arms and ankles from bites after dark. Yellow fever is an even more deadly disease, but immunization is easily obtainable and gives complete protection.

Schistosomiasis, another serious disease, is caused by a worm carried by a species of freshwater snail. Since rivers in Africa and some other parts of the tropics are often infected with the disease, it is wise to consult with local medical authorities before swimming in any jungle river.

## *Finding your way*

In the interior of the jungle, it is seldom easy to find your way around. Good maps, moreover, are rarely available in jungle countries. Thus, in order to know where you are, it is absolutely essential to carry a compass at all times. If you are traveling where there is no trail, mark your route by blazing large trees with a machete or by bending down or cutting off small saplings along your path.

If you are planning a long expedition, be sure to inform two or three responsible people beforehand when and where you are entering the jungle, where you are planning to camp, and when you expect to return. If you should get lost or have an accident, do not panic or wander aimlessly. Find an open area where you will be visible to searchers traveling by plane or on foot and *stay there.*

COMPASS

But the most important rule of all is to be sensible in the jungle, just as you would in any wilderness area. If you are properly equipped and have planned your trip well, you will find that the jungle is a place where you can live comfortably and enjoy yourself. The rewards of jungle camping are so many, in fact, that once you have tried it you will want to return again and again.

# How to Make an Insect Collection

In tropical jungles, insects far outnumber all other forms of animal life. They live in the air, in water, in the soil, and on and even inside plants. They vary in size from almost microscopic flies to giant beetles and butterflies, and occur in an astonishing variety of shapes and colors. As a result, collecting tropical insects makes a rewarding and exciting hobby. If they are properly prepared and stored in a dark dry place, specimens will last for years, and you can add to your collection wherever you travel.

## *How to capture specimens*

At ground level, the best places to seek jungle insects are around blossoming plants (especially in sunny clearings), under rotten logs, in soil and among leaf litter, and, at night, near any kind of light source. The upper layers of the forest canopy can yield even more interesting specimens. But, since collecting here is hardly possible without the use of pulleys, rope ladders, and other special equipment, it is best left to professional scientists.

The simplest collecting tool is a standard sweep net. The net consists of nylon mesh or a similar material attached to a wire hoop twelve to fifteen inches in diameter and fastened to a pole about three feet long. Simply sweep the net through the air to catch flying insects such as butterflies. At the end of each stroke, twist your wrist so that the net falls over the opening of the hoop to prevent the escape of any captives. Sturdier nets of muslin or canvas can be swept through foliage or dragged in water to catch other types of insects.

Another collecting method is to spread a sheet beneath plants and then beat the foliage with a stick. Dislodged insects will fall on the sheet. By painting trees or special boards with boiled mixtures of sugar or molasses and rum or alcohol—or with special compounds sold by biological supply houses—you can attract and capture many insects such as night-flying moths.

To capture ground-dwelling insects, use pitfall traps. Bury a tin can or open jar in the ground so that the top is even with the surface and bait it with old meat, fish, fruit, or honey. A slab of wood or flat rock propped over the top with a few small rocks will keep the trap from filling with water when it rains.

PITFALL TRAP

Some of your most exciting catches can be made at night when literally millions of insects are on the wing. Any artificial light will attract an amazing variety of insects, especially on dark moonless nights, although the numbers and species will vary from night to night and from hour to hour, depending on conditions such as weather and the amount of wind. Just as night-flying insects in a city are attracted to neon signs and street lights, jungle insects flock to any bright artificial light. Even as you sit in camp eating your supper, countless insects will fly into your lamp and then fall down on the table or into your food. It is also worthwhile to hang a strong light such as a Coleman lantern or a mercury-vapor lamp about five feet above the ground at the edge of the camp clearing. If a white sheet is hung about a foot behind the lantern and extended forward beneath it, night-flying insects of every kind will alight on the sheet and can easily be collected.

KILLING JAR

## *Killing method*

All specimens you collect must be killed before they can be preserved, mounted, and transported. Large butterflies and moths can be killed by pinching the body between the fingers. Large beetles can be dropped briefly into hot water. Aquatic insects can be put directly into a seventy percent solution of isopropyl alcohol. But for the majority of insects it is best to use a killing jar. You can buy one from a biological supply house, or you can make one yourself.

The safest type for amateurs is a wide-mouthed jar with a screw-top cap. Put a layer of cotton in the bottom of the jar, moisten it with ethyl acetate (if available) or with chloroform (which can be purchased from druggists in most states) and cover the cotton with a round piece of blotting paper to keep the insects from direct contact with the wet cotton. The fumes in the jar will kill most specimens within a few minutes. To keep the insects from drying out, do not leave them in the jar for more than a hour or two. Wrapping the jar with adhesive tape will reduce the possibility of breakage.

## *Labeling specimens*

Always record data for labels as soon as the insects are captured, since unlabeled or improperly labeled specimens have no real value for later identification and study. As a bare minimum, you should record the place where each specimen was collected, date and hour of collection, the collector's name, and the specimen's name (if known). This should be written in India ink on a small label made from a piece of stiff white index card and kept permanently with the specimen. You will enhance the value of your collection if you record additional information about each specimen, such as: the kind of habitat where it lived;

whether you caught it among leaves, under a rock, or wherever; weather conditions when you caught it, and so on.

## *Preserving and mounting specimens*

Soft-bodied insects such as aphids and lice, and most larvae and nymphs, should be stored in small, labeled vials containing a seventy percent solution of isopropyl alcohol. Adding a small amount of glycerin will prevent them from drying out if the alcohol evaporates.

Hard-bodied insects usually are best mounted at home rather than in the field. Before they have a chance to become hard and brittle, the specimens should be packed in small boxes between layers of cellulose cotton, if available. Do not use ordinary cotton, since the insects become entangled in it. Butterflies and moths are best kept in small individual envelopes. The boxes of specimens should then be gently dried over a small fire or stove to prevent the growth of mold. When dry, add some naphthalene (moth flakes) or paradichlorobenzene to discourage insect pests, and keep each box in a sealed plastic bag with silica gel to absorb any moisture. If you are traveling in a foreign country, specimens packed in this way can be mailed home safely at low cost. Label the boxes "dried insects for scientific study—no commercial value."

By the time you are ready to mount the dried specimens, they probably will have become hard and brittle. Relax them by placing them in a metal box that closes tightly. A moist blotter at the bottom of the box will make the air inside humid, and the insects will be pliable enough to work with in twelve to twenty-four hours.

All but the very smallest insects can be mounted on insect pins, which are rustproof and available in a number of sizes. The pin should be passed through the insect's body and then through the identifying label. The insect is stored by pinning it to the bottom of a sturdy box lined with a layer of cork, balsa wood, or styrofoam. A few moth flakes will discourage pests.

Butterflies and moths should be pinned on a spreading board. This consists of two soft wooden boards on which to spread the insect's wings, with a groove between for the body. Position the specimen carefully with pins, and allow it to dry for a day or two before adding it to your permanent collection.

SPREADING BOARD

Very small insects can be stuck with a drop of shellac or other adhesive to the tips of small triangular points cut from stiff paper. An insect pin is then pushed through the base of the paper point, a label is added, and the insect is stored in the usual way.

# Glossary

**Adaptation:** An inherited structural, functional, or behavioral characteristic that improves an organism's chances for survival in a particular *habitat.*

**Aerial root:** A root that normally grows entirely or partly above ground, such as the roots of *epiphytes, prop roots,* or clinging roots growing from the stems of vines.

**Air plant:** *See* Epiphyte.

**Arboreal:** Living in trees rather than on the ground or in air or water. *See also* Terrestrial.

**Aroid:** A plant belonging to the arum family; characterized by the production of flowers on a fleshy spike surrounded by a hoodlike bract.

**Biological nomads:** Fast-growing, short-lived plants, including trees such as musangas and cecropias, which are among the first species to colonize natural or man-made clearings in the jungle. Typically, these trees have seeds adapted for long-distance dispersal and, their seedlings are able to grow in bright sunlight.

**Biomass:** The total weight of of living matter, or of a specific kind of plant or animal, in a given area at a given time. Thus we may refer to the total biomass of a given area of jungle, or to the biomass of the trees, predators, or snakes in an area of jungle.

**Broadleaf:** Term describing a tree with wide-bladed leaves, such as an oak or a maple; generally refers to flowering trees in contrast to conifers, which usually have needle-shaped or scalelike leaves. In temperate forests, most broadleaf trees are *deciduous,* but in jungles, most broadleaf trees are *evergreen. See also* Needleleaf.

**Bromeliad:** A plant belonging to the pineapple family. Although one species is found in Africa, most bromeliads are native to the New World, especially in tropical and subtropical areas. Many live as *epiphytes.*

**Buttress:** A thin, triangular plate of wood in the angle between the trunk and a horizontal root of a jungle tree. Several buttresses usually radiate from the base of a single tree. They are a common feature of many large jungle trees, but are seldom found on trees growing outside the tropics.

**Canopy:** The overhead layer of branches and leaves in a forest.

**Carnivore:** An animal, such as a jaguar, hawk, or praying mantis, that lives by eating the flesh of other animals. *See also* Herbivore; Omnivore.

**Cauliflorous plant:** A plant, such as the cacao tree, that produces flowers and fruits directly on its main trunk or larger branches instead of on smaller twigs. Although many jungle trees and vines are cauliflorous, plants producing blossoms in this way are rarely found outside the tropics.

**Cellular respiration:** The process by which energy is released through the oxidation of organic compounds within living cells of plants and animals. Energy released in this way is used to maintain the organism's life processes.

**Chlorophyll:** A group of pigments responsible for the green color of plants and essential to *photosynthesis.*

**Climate:** The average *weather* conditions of an area, including temperature, rainfall, humidity, windiness, and hours of sunlight, based on records kept for many years.

**Climax community:** The final or mature association of living things in a natural *succession;* the relatively stable, self-sustaining association of living things best suited for survival under existing conditions of soil, climate, and other environmental factors.

**Cloud forest:** A luxuriant forest growing on mountain slopes as a result of persistent clouds, fog, and frequent rain throughout the year. The high level of cloudiness and precipitation is caused by the cooling of moisture-laden air currents when they are deflected upward by the mountain slopes.

**Community:** All the plants and animals living together in a particular *habitat* and bound together by *food chains* and other interrelations.

**Competition:** The struggle between individuals or groups of living things for such common necessities as food, water, or living space.

**Conservation:** The use of natural resources in a way that ensures their continuing availability to future generations; the wise use of natural resources.

**Consumer:** Any living thing that is unable to manufacture food from nonliving substances but depends instead on the energy and nutrients stored in other living things. *See also* Carnivore; Decomposer; Herbivore; Omnivore; Producer.

**Deciduous:** Term describing a plant that periodically loses all its leaves at a certain season each year. This phenomenon is common in temperate forests, but rare in *tropical rain forests,* where most trees are *evergreens.* An individual leaf on a deciduous tree has a life of twelve months or less, while an evergreen leaf remains alive for two years or more.

**Decomposer:** A living organism, such as certain fungi, bacteria, and small animals, that feeds on dead material and causes its chemical and mechanical breakdown. In the process, decomposers obtain energy for their own life processes and release chemical nutrients into the soil.

**Diapause:** A period of dormancy and arrested growth in the life cycles of certain insects. Animals in this state are highly resistant to unfavorable external conditions such as prolonged periods of drought or cold weather. *See also* Hibernation.

**Diurnal:** Active mainly during the day. *See also* Nocturnal.

**Ecology:** The science which studies the relationships of living things to each other and to their nonliving environment. A scientist who analyzes these relationships is an ecologist.

**Ecosystem:** An assemblage of living organisms and their nonliving environment which interacts and is linked by energy and nutrient flow. An ecosystem may be an area as large as a forest or as small as a pond or a square yard of grassland.

**Environment:** All the external conditions, such as soil, water, air, and organisms, surrounding a living thing.

**Epiphyte:** A plant that grows upon or is attached to another plant but, except for mistletoes, usually derives no sustenance from its support. Epiphytes, often called *air plants,* are a characteristic feature of jungle vegetation but also are found outside jungles.

**Equator:** An imaginary circle around the middle of the earth, equally distant at all points from both the North and South Poles.

**Evergreen:** A tree that does not lose all its leaves at one time. In temperate forests most *needleleaf* trees, such as pines and spruces, are evergreens, as are a few *broadleaf* trees, such as live oaks and hollies. In jungles, on the other hand, most trees are broadleaf evergreens. *See also* Deciduous.

**Evolution:** The process by which modern plants and animals have arisen from forms that lived in the past, as a result of gradual change in their inherited makeup.

**Food chain:** A series of plants and animals linked by their food relationships. Green plants, plant-eating insects, and insect-eating birds would form a simple food chain.

**Fungi** (singular *fungus*): A group of plants lacking *chlorophyll*, roots, stems, and leaves. Fungi obtain their food either as *parasites* on living plants and animals or by breaking down dead organic material. Many fungi are important as *decomposers*.

**Habitat:** The immediate surroundings (living place) of a plant or animal; everything necessary to life in a particular location except the life itself.

**Herb:** A flowering plant or fern that has a soft, rather than woody, stem.

**Herbaceous:** Having a soft rather than a woody stem.

**Herbivore:** An animal that eats plants. *See also* Food chain; Omnivore.

**Hibernation:** A prolonged sleeplike state that enables an animal to survive during the winter months in a cold climate. The heartbeat, breathing, and other body processes of the hibernating animal slow down drastically, and it neither eats nor drinks. *See also* Diapause.

**Host:** A living organism whose body supplies food or living space for another organism. *See also* Hyperparasite; Parasite.

**Hyperparasite:** A *parasite* whose *host* is another parasite.

**Jungle:** A word of Indian origin now popularly used for almost any type of tropical forest; most commonly refers to *tropical rain forest* but frequently used to refer to *seasonal forests* of the tropics as well.

**Larva** (plural *larvae*): An active immature stage in an animal's life history, during which its form differs from that of the adult, such as the caterpillar stage in the development of a butterfly. *See also* Pupa.

**Latitude:** Distance north or south of the equator, measured in degrees.

**Liana:** A climbing woody plant that is rooted in the soil but depends on trees or other plants for support. The great abundance of lianas is a characteristic feature of *tropical rain forests*, where these woody vines often have stems of great length and thickness.

**Machete:** A large, heavy-bladed knife used as a tool for cutting brush, harvesting crops such as sugar cane, or as a weapon; called a cutlass in the West Indies, a parang in the East Indies, and a bush knife in West Africa.

**Mammals:** The group of animals including humans, bats, *rodents*, and many other forms. All are warm-blooded, possess special milk-producing glands, are at least partially covered by hair, and usually bear their young alive.

**Microclimate:** The *climate* or sum of environmental conditions such as temperature, humidity, and air movement, in a very restricted area or space, such as on the jungle floor or inside a hole in a tree. The microclimate of a small space may differ considerably from the climate of the surrounding area.

**Microhabitat:** A miniature *habitat* within a larger one; a restricted area where environmental conditions differ from those in the surrounding area. The water tank of a bromeliad is an example of a microhabitat in the jungle.

**Mycelia:** The threadlike filaments that compose the plant body of a *fungus*.

**Mycorrhiza:** The close association of a *fun-*

*gus* with the roots of a higher plant. It is believed that in some cases mycorrhiza may benefit plants by enabling the fungus to transfer mineral nutrients from decomposing organic matter directly to the roots of plants.

**Myrmecophyte:** An "ant plant"; a plant that is normally inhabited by ants. The reasons why myrmecophytes are found only in the *tropics* are uncertain.

**Needleleaf:** Bearing needlelike leaves; usually refers to coniferous trees such as pines and spruces. *See also* Broadleaf.

**Niche:** An organism's role, or "occupation," in a natural community, such as *scavenger* or nocturnal *predator*. The term refers to the organism's function, not the place where it is found.

**Nocturnal:** Active mainly at night. *See also* Diurnal.

**Omnivore:** A mixed feeder; an animal whose normal diet includes both plants and animals. *See also* Carnivore; Herbivore.

**Parasite:** A plant or animal that lives in or on another living thing (its *host*) and obtains part or all its food from the host's body, but usually without killing its host. *See also* Hyperparasite.

**Photosynthesis:** The process by which green plants use carbon dioxide and water to make simple sugars. *Chlorophyll* and sunlight are essential to the series of complex chemical reactions involved in the process.

**Pollination:** The transfer of pollen from the male to the female organs of flowers, resulting in the formation of seeds.

**Predator:** An animal that lives by capturing other animals for food.

**Prehensile:** Adapted for grasping, seizing, or holding on, such as the tails of certain New World monkeys.

**Prey:** A living animal that is captured for food by another animal. *See also* Predator.

**Producers:** Green plants, the basic link in any *food chain*. By means of *photosynthesis*, green plants produce the food on which all other living things ultimately depend. *See also* Consumer.

**Productivity:** The total mass of organic matter produced over a given period of time in a functioning natural community, by a group of plants, or by an individual plant.

**Profile diagram:** A diagram of a forest as seen in profile view, based on accurate measurements and drawn to scale. The diagram records the species and position of each tree present, as well as such things as their heights, branching patterns, and the diameters of their trunks and crowns.

**Prop root:** An exposed root that curves downward from a position near the base of the main trunk of a tree. Prop roots, also called stilt roots, are common on many small to medium jungle trees. They are also found on several kinds of trees that live outside the jungle, as well as on some herbaceous plants, such as corn.

**Pseudobulb:** A bulblike enlargement at the base of the stem, found on many species of epiphytic orchids. During dry periods, the plants draw on reserves of water stored in the tissues of pseudobulbs.

**Pupa** (plural *pupae*): The relatively inactive stage in the development of certain insects during which the *larva* transforms into an adult, such as the chrysalis stage in the development of a butterfly.

**Rodents:** The large group of *mammals* characterized by possession of continuously growing front teeth adapted for gnawing. Rats, mice, squirrels, and beavers are typical examples of rodents.

**Scavenger:** An animal that eats the dead remains and wastes of other animals and plants.

**Seasonal forest:** A tropical forest which, in

contrast to the ever-wet *tropical rain forest,* has an annual dry season of several months. In such forests, many of the trees shed all their leaves in the dry season.

**Shifting cultivation:** A method of farming, also known as "slash-and-burn" agriculture, in which clearings are made in the forest, crops are planted for one or a few growing seasons, and then the plots are abandoned and new clearings are made and planted.

**Species** (singular or plural): A group of plants or animals with many characteristics in common. Individuals belonging to the same species resemble each other more closely than they resemble individuals of any other species and usually interbreed only with each other.

**Stratification:** The arrangement of the crowns of different species of trees in a forest into more or less orderly horizontal layers. *See also* Profile diagram.

**Succession:** The process of continuous, gradual replacement of one community of plants and animals by another over a period of time, eventually leading to a more or less stable *climax community.*

**Terrestrial:** Living on the ground rather than in air, water, or trees. *See also* Arboreal.

**Tropic of Cancer:** An imaginary circle around the earth 23.45 degrees north of the *equator.* It forms the northern boundary of the *tropics.*

**Tropic of Capricorn:** An imaginary circle around the earth 23.45 degrees south of the *equator.* It forms the southern boundary of the *tropics.*

**Tropical rain forest:** The natural *climax* vegetational *community* of equatorial lowlands. The almost completely nonseasonal *evergreen* forest of *tropical climates* with abundant rainfall well distributed throughout the year; characterized by the presence of many species of tall trees and an abundance of thick-stemmed *lianas* and woody and herbaceous *epiphytes.* Popularly known as *jungle,* although this term is also used in reference to tropical *seasonal forests.*

**Tropics:** The area of the earth's surface lying between the *Tropic of Cancer* and the *Tropic of Capricorn.* Characterized by the absence of winter and, except at high altitudes, by constantly warm temperatures.

**Weather:** The condition of the atmosphere over a relatively short period of time, in terms of temperature, humidity, windiness, presence or absence of precipitation, and clearness or cloudiness. *See also* Climate.

# Bibliography


**JUNGLES AND JUNGLE LIFE**

ALLEN, PAUL H. *The Rain Forests of Golfo Dulce.* University of Florida Press, 1956.

AUBERT DE LA RÜE, EDGAR, FRANÇOIS BOURLIÈRE, and JEAN-PAUL HARROY. *The Tropics.* Knopf, 1957.

BATES, MARSTON. *Where Winter Never Comes.* Scribner, 1952.

BATES, MARSTON, and THE EDITORS OF LIFE. *The Land and Wildlife of South America.* Time, Inc., 1964.

BROWN, LESLIE. *Africa: A Natural History.* Random House, 1965.

CARR, ARCHIE, and THE EDITORS OF LIFE. *The Land and Wildlife of Africa.* Time, Inc., 1964.

COLLINS, WILLIAM B. *The Perpetual Forest.* Lippincott, 1959.

DORST, JEAN. *South America and Central America: A Natural History.* Random House, 1967.

PFEFFER, PIERRE. *Asia: A Natural History.* Random House, 1968.

RICHARDS, PAUL W. *The Tropical Rain Forest.* Cambridge University Press, 1952.

RIPLEY, S. DILLON, and THE EDITORS OF LIFE. *The Land and Wildlife of Tropical Asia.* Time, Inc., 1964.

SILVERBERG, ROBERT. *The World of the Rain Forest.* Meredith, 1967.

WALLACE, ALFRED R. *The Malay Archipelago.* Dover, 1962.

**ECOLOGY**

ALLEE, W. C., and KARL P. SCHMIDT. *Ecological Animal Geography.* Wiley, 1951.

BATES, MARSTON. *The Forest and the Sea.* Random House, 1960.

OWEN, DENNIS F. *Animal Ecology in Tropical Africa.* Oliver and Boyd, 1966.

SCHULZ, JOHAN P. *Ecological Studies on Rain Forest in Northern Suriname.* North-Holland, 1960.

**JUNGLE PLANTS**

CORNER, ELDRED J. H. *The Natural History of Palms.* Weidenfeld and Nicholson, 1966.

CORNER, ELDRED J. H. *Wayside Trees of Malaya.* Government Printing House of Singapore, 1952.

HOLTTUM, RICHARD E. *Plant Life in Malaya.* Longmans, Green, 1954.

MORS, W. B., and C. T. RIZZINI. *Useful Plants of Brazil.* Holden-Day, 1966.

MORTON, JOHN K. *West African Lilies and Orchids.* Longmans, Green, 1961.

PURSEGLOVE, JOHN W. *Tropical Crops: Dicotyledons.* Wiley, 1968.

WILLIAMS, L. O. *Tropical American Plants II.* Field Museum of Natural History, 1961.

**BIRDS**

BANNERMAN, DAVID A. *The Birds of West and Equatorial Africa.* Oliver and Boyd, 1953.

BLAKE, EMMET R. *Birds of Mexico.* University of Chicago Press, 1953.

CHAPIN, JAMES P. *The Birds of the Belgian Congo.* Bulletin of the American Museum of Natural History, Vols. 65, 75, 75A, 75B (1932–1954).

DE SCHAUENSEE, R. MEYER DE. *The Birds of Colombia.* Livingston, 1964.

ELGOOD, JOHN H. *Birds of the West African Town and Garden.* Longmans, Green, 1960.

HAVERSCHMIDT, FRANÇOIS. *Birds of Surinam.* Oliver and Boyd, 1968.

MOREAU, R. E. *The Bird Faunas of Africa and Its Islands.* Academic Press, 1966.

SLUD, PAUL. *The Birds of Costa Rica.* Bulletin of the American Museum of Natural History, Vol. 128 (1964).

SMITHE, FRANK B. *The Birds of Tikal.* Natural History Press, 1966.

SMYTHIES, BERTRAM E. *The Birds of Borneo.* Oliver and Boyd, 1968.

SMYTHIES, BERTRAM E. *The Birds of Burma.* Oliver and Boyd, 1953.

**OTHER ANIMALS**

BATES, MARSTON. *The Natural Histories of Mosquitoes.* Macmillan, 1949.

BEEBE, WILLIAM, G. INNESS HARTLEY, and PAUL G. HOWES. *Tropical Wildlife in British Guiana.* New York Zoological Society, 1917.

BOOTH, ANGUS H. *Small Mammals of West Africa.* Longmans, Green, 1960.

CANSDALE, GEORGE S. *Reptiles of West Africa.* Penguin, 1955.

GEE, EDWARD P. *The Wild Life of India.* Collins, 1964.

KLOTS, ALEXANDER B., and ELSIE B. KLOTS. *Living Insects of the World.* Doubleday, 1959.

LEOPOLD, ALDO S. *Wildlife of Mexico.* University of California Press, 1959.

RICHARDS, OWAIN W. *The Social Insects.* MacDonald, 1953.

TWEEDIE, MICHAEL W. F. *The Snakes of Malaya.* Government Printing House of Singapore, 1957.

**JUNGLE TRAVELERS**

BATES, HENRY W. *Naturalist on the River Amazons.* University of California Press, 1962.

BEDDALL, BARBARA G. *Wallace and Bates in the Tropics.* Macmillan, 1969.

BEEBE, CHARLES W. *High Jungle.* Duell, Sloan, and Pearce, 1949.

BEEBE, CHARLES W. *Jungle Days.* Putnam, 1925.

BEEBE, CHARLES W. *Jungle Peace.* Holt, 1919.

CARR, ARCHIE. *High Jungles and Low.* University of Florida Press, 1953.

HINGSTON, RICHARD W. G. *A Naturalist in the Guiana Forest.* Longmans, Green, 1932.

MATTHIESSEN, PETER. *The Cloud Forest.* Viking, 1961.

# *Illustration Credits and Acknowledgments*

**COVER:** Spider monkey, Karl Weidmann

**ENDPAPERS:** Carl Frank

**UNCAPTIONED PHOTOGRAPHS:** *8–9*: Jungle scene, Barro Colorado, David Cavagnaro *68–69*: Jungle riverbank, South America, Karl Weidmann *118–119*: Coatimundi, South America, Alan Root *158–159*: Waika tribesman, South America, Karl Weidmann

**ALL OTHER ILLUSTRATIONS:** *10–11*: Alan Root *12–13*: Karl Weidmann *14*: Photo Library *15–16*: Paul W. Richards *17*: J. Allan Cash *18*: The Equatorial Jungle, Henri Rousseau, National Gallery of Art, Washington, D.C., Chester Dale Collection *20*: Karl Weidmann *21*: Edward S. Ross *22*: Nick Drahos *23*: Nick Drahos; F. Hall, Photo Researchers *24–25*: Alan Root *26*: Harald Schultz *27*: Allan Roberts *28–29*: Harald Schultz *30*: Alan Root *32–33*: Charles Fracé *34*: Roy Pinney, Photo Library *36–37*: Karl Weidmann *38*: Lorus and Margery Milne *39*: Nick Drahos *40–41*: Karl J. Frogner *42*: Lorus and Margery Milne *42–43*: Robert S. Simmons *45*: M. P. L. Fogden *46–47*: Patricia C. Henrichs *48*: Karl Weidmann *49*: M. P. L. Fogden *51*: Grant Haist *52*: M. Krishnan *53*: Ivan Polunin *55*: B. Brower Hall *56–57*: Graphic Arts International *58–59*: Cecilia Duray-Bito *60*: Hans Zillessen, G.A.I. *61*: Emil Javorsky *62–63*: Nick Drahos *64–65*: Emil Javorsky *66*: Peter Ward *70*: Lord Medway *71*: G. Holton, Photo Researchers *72*: M. Krishnan *74*: Alan Root *75*: Douglas Faulkner *77*: James A. Kern *78*: Karl J. Frogner *79*: Edward S. Ross *80*: Karl J. Frogner *81*: Ivan Polunin; Peter Ward *82*: Dale A. Zimmerman *83*: David Chivers *84*: James A. Kern *86*: Paul Popper Limited *87*: M. P. L. Fogden *88–89*: Patricia C. Henrichs *90–91*: Hans Zillessen, G.A.I. *92*: Weldon King *93*: Alan Root *94–95*: Karl Weidmann *96*: Cecilia Duray-Bito *97*: Charles Fracé *98*: Jen and Des Bartlett *99*: J. Simon, Photo Researchers *101*: Karl Weidmann *102*: M. P. L. Fogden *104–105*: Carl W. Rettenmeyer *106*: Patricia C. Henrichs *107*: E. Aubert de La Rue *108*: E. R. Degginger *109*: Patricia C. Henrichs *110*: Walter Dawn *111*: Tony Beamish; J. Jangoux, Photo Researchers *112*: Emil Javorsky; E. R. Degginger *113*: Thase Daniel *115*: Ivan Polunin *116*: Karl Weidmann *120–121*: Ann and Myron Sutton *122–123*: Graphic Arts International *124*: John H. Kaufmann *125*: Thomas Eisner *126*: Edward S. Ross *127*: M. P. L. Fogden *128*: James A. Kern *130*: Alan Root *131*: Charles Fracé *132–133*: Karl Weidmann, National Audubon Society *134–139*: Edward S. Ross *140*: Cecilia Duray-Bito *141*: Patricia C. Henrichs *142*: Karl Weidmann *145–146*: Edward S. Ross *147*: B. Brower Hall *148*: Nick Drahos *149*: Karl Weidmann *150–151*: Alan Root *152–153*: Ivan Polunin *154*: Daniel Janzen *155*: Don Veirs; Carl W. Rettenmeyer *156*: Alan Root *161–162*: L. Renault, Photo Researchers *164–166*: Ivan Polunin *168–169*: Karl Weidmann, National Audubon Society *170*: Karl Weidmann *171*: Ruth Smiley *172*: Karl Weidmann *173*: Grant Haist; David G. Allen *174*: Thase Daniel; Charles F. Leck *176–177*: Cecilia Duray-Bito *178–179*: G. R. Roberts, Nelson, New Zealand *180*: Hans Zillessen, G.A.I. *181*: Edward S. Ross *182–183*: Karl Weidmann *185*: Jen and Des Bartlett *186*: Patricia C. Henrichs *187*: Mann, Monkmeyer Press Photo Service *188–189*: Hans Zillessen, G.A.I. *190*: Carl Frank *191*: Edward S. Ross *192*: Don Eckelberry *193*: Bernheim, Rapho-Guillumette *195–197*: J. Allan Cash *198*: Karl Weidmann *201*: Cecilia Duray-Bito *202–214*: Charles Fracé *215–221*: Graphic Arts International

**PHOTO EDITOR:** BARBARA KNOWLTON

**ACKNOWLEDGMENTS:** *The author wishes to thank Roger A. Beaver and J. M. Cherrett of the University College of North Wales, both of whom assisted in various ways in preparation of this book. Of the many people who gave helpful advice to the editors, special thanks are due to Dean Amadon, Alice Gray, and John C. Pallister of the American Museum of Natural History and Wayne King of the New York Zoological Society. The editors are also grateful to C. Gordon Fredine and William L. Perry of the National Park Service for reading the entire manuscript and offering many useful suggestions.*

# Index

Page Numbers in **boldface** type indicate reference to illustrations.

Javan rhinoceros, **207**, 209